THE GOD OF MONKEY SCIENCE

THE GOD OF MONKEY SCIENCE

People of Faith in a Modern Scientific World

JANET KELLOGG RAY

WILLIAM B. EERDMANS PUBLISHING COMPANY
GRAND RAPIDS, MICHIGAN

Wm. B. Eerdmans Publishing Co.
4035 Park East Court SE, Grand Rapids, Michigan 49546
www.eerdmans.com

Published 2023

Book design by Lydia Hall

Printed in the United States of America

29 28 27 26 25 24 23 1 2 3 4 5 6 7

ISBN 978-0-8028-8319-3

Library of Congress Cataloging-in-Publication Data

A catalog record for this book is available from the Library of Congress.

For my dad, Kirby Kellogg,
who asks the hard questions,
and in memory of my precious mom

Contents

Acknowledgments

It's hard work writing and editing a book and launching it into the world, and I love every minute of it. The people in my corner of the book world are simply the best, and there just aren't enough thank-yous.

To my husband Mark—you're a brainiac and a science geek and a lover of adventures, exactly the partner I need. You urge me to use my voice and be true to my message. You're my best cheerleader, publicist, front-row fan, and tech guy.

To Tommy and Melinda Ballard for reading rough drafts and especially for the wisdom and theological insights from the Ballard School of Divinity. Life friends are the best.

To Sheila Legan, my dear friend of many talents, for reading early drafts of this book and spotting mistakes with the precision of the Webb telescope.

To my Matters More book study group—what a windfall to find myself in this group of women, bound together by our commitment to love above all. As always, you are a safe place for the hard questions.

To Allan Chapman, man of science and man of faith, legendary science writer, Britain's national treasure, and my friend. You inspire me.

I'm grateful for the love and support of my two adult kids, Tabitha and Austin—you are smart and funny and all-around amazing humans.

To the Eerdmans team, talented all, but excelling in support and encouragement. Thanks especially to project editor extraordinaire Laurel Draper for once again providing both expertise and hand-holding. A million thanks for the guidance of acquisitions editor Trevor Thompson and for the talents of marketing director Sarah Gombis. And to the Eerdmans art department, you knocked it out of the park for a second time with cover art design.

1

The Playbook

My family went to church three times a week—twice on Sunday, and again on Wednesday night. During the annual "gospel meeting" (what some traditions call "revivals"), we went to church every single night, for an entire week.

Probably the most scarring aspect of this upbringing was the fact that I never saw *The Wizard of Oz* until I was an adult. In the ancient days before we could record television programs and *long* before on-demand programming, *The Wizard of Oz* was broadcast only on Sunday nights. I managed to see the *Wonderful World of Disney* (a weekly Sunday-night program) a handful of times when I was fortunate enough to be home sick.

In the tradition of my growing-up years, we took church attendance very seriously. We took doctrine seriously. We took the Bible seriously, and of course, we took it literally.

Evolution was a nonissue in my church. In my evangelical tradition, we would have no more questioned the Genesis creation story than we would have questioned the existence of Jesus.

But I fell in love with science in school and then in college, and now I teach biology at a large state university. When I was a young adult, rejection of the evidence for evolution and the evidence for an ancient earth began to feel intellectually dishonest.

I realized it didn't have to be that way. I realized that science and faith were answering different questions.

I read. I studied. I wrote. I explored Christian rejection of evolution, and I looked at all the ways people of faith attempt to force-fit modern science into Genesis. I wrote a book about it.[1]

My book was in the final phases of editing as the COVID pandemic reached the boiling point. I also blog about science, faith, and culture, so all the while I was preparing for the launch of the evolution book, I was writing about the virus and vaccine development and the science of the pandemic.

In the excitement and hubbub of bringing my first book into the world, there was an unavoidable background noise of all things COVID, and particularly, the response of evangelical Christians during the pandemic:

- White evangelicals consistently lagged behind all other religious demographic groups in getting a COVID vaccine.
- Evangelical churches were demanding to meet in person, at the height of the prevaccine pandemic.
- Evangelicals were vocally antimask.
- Evangelical Christians were the loudest voices in some of the ugliest criticisms against leading scientists and epidemiologists.

Of course, evangelicals were not the only ones resisting vaccines, masks, and social distancing, but evangelicals were by far the largest religious demographic doing so.

Why was the evangelical precinct of Christianity the most resistant to masks, vaccines, and even the particulars of the science of the virus and the pandemic itself? It was a puzzle.

In the spring of 2021, when less than 7 percent of the state's population was vaccinated, the governor of Texas lifted the mask mandate, setting off a firestorm of opposing opinions. While some Texans hurled their masks into bonfires in celebration, others warned of continuing community spread with vaccination rates still in single digits.

Within this context, I wrote about the lifting of the mask mandate. I wasn't for endless masking; I was just questioning the timing.

Well, as it happens when you write something for others to read . . . a reader disagreed with my position. I'll spare you the entire unadorned response, but here's the abbreviated version:

> There she goes again . . .
> Janet and her monkey god science.[2]

What?

The comment obviously wasn't meant as a compliment to my astute analysis of the situation. It was also obvious that it was a reference to the writing I do about evolution acceptance and faith.

The description of my masking position as "monkey god science" was not really offensive—it was more perplexing than it was anything else. After all, I wasn't talking about the science of evolution. I was talking about medicine, vaccines, and public health policy.

What did COVID precautions have to do with my accepting evolution science? Apples and oranges, it seemed to me. Monkeys and bonobos. Pick your contrast.

Thinking in Circles

The more I pondered, the more I thought in circles. Not meandering, unrelated circles, but the overlapping circles of a classic Venn diagram.

In a circle representing evangelicals who deny the science of evolution and a circle representing evangelicals who deny the science of all things COVID, I saw a sizable area of overlap. And dropping down into the mix was a circle of evangelicals who deny human-caused climate change.

In the mix are three very different areas of science—biology, epidemiology, and earth science. Why the overlap of denial?

The reader who took issue with my "monkey god science" wasn't alone. David Croom has lots of hot takes on the COVID pandemic which he regularly tweets to his seventeen thousand followers. Here's one from September 2021: "The same people who invented the vaccine believe humans evolved from apes!"[3] If only he'd added the hashtag #MonkeyScience.

Invariably, says climate scientist Katharine Hayhoe, those who post criticisms on her social media are anti-masking on theirs.

I had so many questions.

With very few exceptions, evangelicals embrace science in their lives. Outside small sects who reject modern technology or modern medicine, evangelicals welcome the twenty-first-century life that twenty-first-century science provides.

We aren't denying the science of gravity or the water cycle or photosynthesis. Evangelicals are traveling by plane and car, using smartphones, and taking ibuprofen for headaches.

Evolution, climate science, and COVID appear to be some sort of denialist package deal.

Ground Zero

First, a clarification. When I use the term *science denial*, I'm not talking about incorrect science, although incorrect science troubles me. As a science educator and a Christian, it troubles me when the science for evolution is misrepresented. I am troubled by arguments for creationism that torture both Scripture and science.

Denial is not necessarily a rejection of scientific facts, but rejection of facts can play a role in science denial. We build a straw man, then we tear him down. For example, what most people who reject evolution know about evolution comes from antievolution resources. The evolution that most evangelicals reject seldom resembles the actual evidence for evolution.

But that's a conversation for another day.

Instead, I want to look at how evangelicals talk *about* science. I want to look at how evangelicals talk *about* scientists.

So, I went back to ground zero. I went back to the event that first filled evangelicals with angst about science: the Scopes monkey trial.

Conventional wisdom says, "There's no such thing as bad publicity," and the leadership of Dayton, Tennessee, was happy to volunteer the town as a test case for the adage.

The American Civil Liberties Union (ACLU) wanted to challenge a Tennessee law forbidding the teaching of evolution in schools and advertised for a willing defendant. High school teacher John Scopes volunteered, and town leaders welcomed the opportunity to bring national attention to little Dayton.

The trial took place in 1925 during a hot, pre-air-conditioning summer. The atmosphere was a circus, a carnival, and a parade, all at once. Little girls held monkey dolls bought from vendors and an actual chimpanzee named Joe Mendi, dressed in a dapper suit, sipped a Coca-Cola.

The opposing attorneys were both larger-than-life folk celebrities. Today, they would be hosting highly rated cable news shows.

Scopes was convicted, not a surprise. The ACLU never intended to win. Their goal was an appeal to the United States Supreme Court.

The conviction was overturned, however, and so the case never made it to the Supreme Court.

But no matter—the narrative was cast.

Winning prosecuting attorney William Jennings Bryan focused his arguments on three claims:

- Evolution is not supported by science.
- Evolution undermines morality, society, and religion.
- A fair society would not allow a concept like evolution to be taught in schools.[4]

Darwin's *Origin of Species* had been published sixty-six years earlier. So why now? Why did the Scopes trial put evolution on the map?

Darwin spoke to the scientific community. Folk hero Bryan cast evolution in the language of the people, not of scientists.

The people weren't interested in the scientific evidence. In fact, no scientists testified at the Scopes trial. What the people heard were the consequences of evolution, according to Bryan. They heard about the cost of evolution—to their families, to their faith, to their freedoms. Bryan spelled it out in plain language.

Denialist Playbook

Going forward from the Scopes trial to the present, we see the course Bryan set for evangelical science denial. We see ridicule, disparagement, even vilification of science and scientists.

Anti-evolution sentiment was already brewing in the 1920s. Just three years before Scopes, J. Frank Norris, pastor of the First Baptist Church in Fort Worth, Texas, called up the local zoo and had a load of monkeys delivered to the pulpit. Norris proceeded to interview the monkeys, asking them to "speak for themselves."[5] Working the crowd up into uproarious laughter, Norris launched into a full-throated ridicule of evolution. Live monkeys in the sanctuary was definitely a pastor mic-drop moment, not soon to be equaled. Monkeyshines trump a science lecture, every time.

Fast forward to the twenty-first century and evolution is still good for a clever preacher joke. This time, I had a front-row seat. I was visiting a large church in my hometown and the very popular young minister preached a sermon that can only be described as hot-topic bingo. Evolution was square one. "God will laugh at you when you get to heaven if you believe this nonsense," he declared. The congregation responded with approving laughter and amens.

Monkeys would have been funnier.

And in a 2022 interview with the lead pastor of Sugar Hill Church, the frontrunner for US Senator from Georgia, Herschel Walker (the former NFL star), made this comment: "At one time science said man came from apes, did it not? . . .

If that is true, why are there still apes?"[6] To which the pastor chuckled, "Now Herschel, you're getting too smart for us!"

For decades, evangelical leaders, pastors, and teachers have been telling us that scientists can't be trusted. For decades, evangelicals have been told that scientists are hiding or ignoring evidence for special creation, a young earth, and a global flood.

The "scientists can't be trusted" message is carried by publications, organizations, venues, and ministries, large and small. Some are dedicated to promoting creationism while others have a broader message, with evolution denial promoted under the larger banner. These organizations have podcasts and YouTube channels and websites and in-person events and conferences and *lots* of creationist rafting trips through the Grand Canyon. A tremendous source of antiscience rhetoric comes from popular homeschool and Christian school curricula.

Maybe you've never personally attended an event, watched a documentary, or taken a creationist rafting trip, but if you have spent any time swimming in evangelical waters, you have been exposed to antiscience messages, either directly or indirectly. If not you, your pastor has, or your Sunday school teacher, or your kid's youth pastor, or the youth group volunteers.

If you have spent time in the evangelical world, there are far fewer than six degrees of separation between you and such books, podcasts, videos, movies, blogs, articles, and Christian school curricula. Whether direct or implied, the message is there.

For decades, we accepted all the other sciences, but not the science of evolution. We wanted the march of progress, so we put evolution in its own little time-out box of science.

Evolution was the science that first offended, but in evolution denial, evangelicals developed a whole way of talking about science. Credit the original outline to William Jennings Bryan, but evangelicals took the same talking points and applied them to any science that challenged our theology, our worldview, and our freedoms.

Regardless of the field of science, Bryan's game plan is unmistakable: (1) the scientific evidence is sketchy, misrepresented, or simply wrong, (2) science threatens faith and morality, and (3) acceptance of science comes at a cost to personal freedoms or personal beliefs.

If we believe that scientists are deceiving us regarding evolution, what else are they lying about? If scientists are guilty of hiding evidence for creation or a global flood, are they hiding evidence now for ivermectin or hydroxychloroquine? Are they manipulating climate data for nefarious purposes?

Are scientists in the pockets of Big Pharma? Are scientists just shills for the government?

Evolution-denial arguments are retooled and reloaded and relaunched to serve another day, another topic.

Darwin 2.0

Caroline Matas is a scholar of American evangelicalism and media. In the fall of 2021, Matas attended a three-day conference of the flagship creationist organization, Answers in Genesis, as a researcher-observer.[7]

Matas immediately noticed that she was the only attendee wearing a mask. Although the delta variant was spiking at the time, the only other masks present in the venue were masks sold by vendors—masks emblazoned with antimask messages.

Ken Ham, founder and CEO of Answers in Genesis, has built a creationist empire—the Ark Encounter, the Creation Museum, and the coming Tower of Babel theme park—all his. Answers in Genesis is a powerhouse in creationist literature and media.

Although the organization was born out of the fight against evolution, you are just as likely to find anti-COVID-science and anti-climate-science content from Answers in Genesis.

At the conference, Matas observed a consistent theme: scientists cannot be trusted. Any scientist with a worldview outside the one Answers in Genesis endorses (evangelical, young-earth creationist, literal and historical Genesis) is, by default, wicked and hostile to God.

When looked at through a biblical lens (a young earth and a literal, historical Genesis), scientific evidence will always be interpreted correctly. Any evidence that contradicts a young earth and instantaneous creation is simply being misinterpreted.

For three days, Matas listened to speakers and perused book offerings. The strength of the creationist narrative, noted Matas, is that all conclusions are foregone conclusions, regardless of any contradictory evidence. With foregone conclusions, you'll never be wrong.

There was a time when scientists were the good guys—heroes even—but those days are no more. Edward Humes is the author of *Monkey Girl: Evolution, Education, Religion, and the Battle for America's Soul*, the story of a community torn apart in a battle over teaching evolution in the local public schools. Humes's analysis is unsettling: "Scientists, long the heroes, were now the bad guys. Truth hiders. God killers."[8]

Charles Darwin started the trouble, but he's long gone. Darwin is no longer number one. There's now a new enemy, a new Darwin in town. Scientists are the new Darwin.

It was bad enough when scientists endorsed evolution; now they are preaching to us about germ theory and RNA and carbon dioxide levels, and we are not having it.

Inside the Camp

Evangelicals are my people. I was raised in an evangelical church. Evangelicals introduced me to Jesus. I'm not critiquing evangelicals as an outsider. I want to describe the evangelical relationship with science from inside the camp. I am still in an evangelical church. It is my tradition.

Am I in agreement with everything I hear at church? No, but then I don't agree with myself of twenty years ago.

Am I in agreement with the scientific views of most of the members in my church? Not even close.

It's important to define our terms. What is an *evangelical*? It is not as cut and dried as where you go to church, or even *if* you go to church.

Evolution was the first science field found so offensive by evangelicals that entire organizations were formed to fight it. For the sake of our discussion, how will we consider the theory of evolution?

All evangelicals are not young-earth creationists, and many Christians who accept the evidence for evolution acknowledge God as Creator. The term *creationist* is nuanced.

We need to know where we stand on these terms, so on to the next chapter.

2

Science and the Evangelicals

After driving six miles down gravel roads, you stop and park. All around, prairie grass and wildflowers grow up from land so flat you can see for miles. In the distance, a grouping of seventy-foot-tall, monumental rock formations jut up from the ground, like a prairie-land Stonehenge. We are in western Kansas, and these rock outcroppings are called Monument Rocks, part of the Niobrara Chalk Formation. Nearby is a similar outcropping called Castle Rock.

One hundred million years ago, North America was bisected by the Western Interior Seaway, a shallow sea connecting the Gulf of Mexico with the Arctic Ocean. All of Texas, Colorado, and Kansas were seabed at one point, as were other central and southeastern states. It was a busy few million years—the tectonic plate movement that opened the seaway was also heaving up the Rocky Mountains.

The marine waters teemed with microscopic life—tiny shelled animals called *foraminifera*. When the foraminifera, algae, and other organisms died, they sank to the bottom and accumulated as a type of sediment called *ooze* (yes, that's its real name). The accumulating ooze loosely cemented together in a soft form of limestone called *chalk*. When overlying rock layers eroded, the Niobrara Chalk was exposed at the surface. As the exposed chalk began to erode, some remnants of the chalk were protected when more resistant beds above shielded softer layers below. The uneroded chalk remnants are Monument Rocks and Castle Rock in western Kansas.

The heyday of the Western Interior Seaway, however, was not just a constant, silent rainfall of tiny foraminifera on the sea floor. The seaway was filled with giants.

Looking like a cross between a snake and a shark, massive mosasaurs hunted the seas of western Kansas. Mosasaurs were sharp-toothed reptiles, fearsome predators with awesome maneuverability. *Tylosaurus*, a Kansas native, grew up to forty feet long. Some of the best fossilized mosasaurs in the world once swam the Kansas segment of the seaway.

Mosasaurs weren't the only giant reptiles making Cretaceous Kansas their home. Enormous plesiosaurs, looking like someone stuck a snake on a turtle's body, also swam the seaway. Flying reptiles, the pterosaurs, glided over the seaway looking for food.

Yes, Dorothy, we actually *are* still in Kansas, just a little before your time.

Since the nineteenth century, Kansas has been a fossil hunter's paradise. "In any major museum in this time period, look at the marine Cretaceous stuff and it will be from Kansas," says Rex Buchanan, former director of the Kansas Geological Survey.[1] It's true—I've seen Kansas mosasaurs in New York, Washington, DC, and London.

Kansas is proud of its paleohistory, and rightly so. The Sternberg Museum of Natural History in Hays, Kansas, has many of these giants on display, as well as re-creations of the ancient shoreline with life-size models of the inhabitants.

Visitors, fossil hounds, and paleontologists flock to Kansas. It's a great place to learn about the Western Interior Seaway and its Cretaceous population.

A great place to learn, unless you were a kid in Kansas public schools at the turn of the twenty-first century.[2]

In 1999, the Kansas State Board of Education planned to adopt new science standards. As is usual, the board convened a committee of scientists and science educators to craft a set of standards for the state.

The science committee submitted a set of standards using a nationally recognized curriculum as a model. The mostly evangelical state board of education promptly dumped the proposed standards in favor of a set of standards written by citizen volunteers, directed by a local young-earth-creationist organization.

Gone were any references to cosmology, biology, and geology that might contradict a six-to-ten-thousand-year-old earth and instantaneous special creation. Adding insult to injury, the board even changed the definition of *science* in the standards.

Before you knew it, a history textbook for seventh and eighth graders in Kansas deleted an entire chapter, wiping out all references to the ancient Kansas sea, the fossils found there, and the mosasaurs so prominently displayed in the Sternberg Museum.[3]

Why the mosasaur hate?

First of all, you need long periods of time for chalk formation: millions of years of ooze, millions of years to cement into chalk.

And that's only the beginning. Mosasaurs and plesiosaurs are reptiles whose ancestors walked on land, with legs. The timescale needed for evolutionary adaptations to a marine lifestyle does not fit a six-to-ten-thousand-year young-earth creationist framework.

School kids in Kansas sat in school buildings built on top of a history that didn't happen, according to the science standards in their state.

Creationism Redux

Two years later, elections changed the composition of the Kansas State Board of Education, and evolution and Kansas's Cretaceous history returned to class.

But don't hold your breath, mosasaurs. Three years later, another election, another evangelical board, and again the mosasaurs were kicked to the curb.

This time, however, the state board employed a term favored by some in creationist circles: *intelligent design*. Instead of erasing all references to evolution, students were instructed to challenge evolution. Evolution was to be picked apart and critiqued.

This time around, the state board empowered teachers to discuss the possibility of a supernatural designer *in science class*, as a scientific alternative to biological evolution.[4]

The Discovery Institute provided parents with a list of ten "gotcha" questions with which to coach their children. Any time a teacher tried to teach about millions of years or evolutionary change, students were encouraged to challenge them with a barrage of questions.

I've been a middle school science teacher. It doesn't take much to throw an emergency brake and derail a class of twelve- and thirteen-year-olds.

In Kansas, evangelicals derailed evolution and millions of years of geologic history, despite the evidence literally under their feet.

Who Are the Evangelicals?

Before we look at how evangelicals talk about science, we need to talk about evangelicals.

Attempting to define any group by strict parameters is destined to fail. Rare is the person who checks every box of a group to which they belong.

Instead of defining, I'm going to describe. Taken as a whole, evangelicals will check at least some of the descriptive boxes.

Doctrinally, there are four marks of evangelical belief:[5]

- The Bible is the sole authority for faith and practice. For many evangelicals, this means the Bible is without error in all its claims, including the nature of the physical world.
- Jesus is the only way to salvation. For most evangelicals, this involves a decision to accept Jesus and a "new birth" experience or event.
- Reconciliation with God is through Jesus, not good works.
- Evangelism, or spreading the good news about Jesus, is central to practice.

Evangelicals can also be described in a social sense:

- There is strict conformity to moral and behavioral codes. Sometimes this leads to separation from aspects of culture and primarily associating with like-minded evangelicals.
- An individual's character is the result of personal choices, with little influence from systemic or institutional evils.
- There is an ongoing conflict between religion and culture. We see this in the "culture war" language prevalent in evangelical circles. Evangelicals sometimes respond by erecting a wall between their life at church

and everything else. In the last few decades, evangelicals looked to government and politics as allies in a culture war.

- An anti-intellectual distrust of scholarship permeates evangelical subculture. This looks like a "common sense" approach to biblical interpretation that ignores cultural and contextual aspects of Scripture. Experts are often suspect.
- Celebrity-driven and brand-driven platforms and networks are preferred over traditional institutions.
- Expressions of faith are wedded to expressions of nationalism. This looks like declarations of "God and country" and "America is a Christian nation."

When pollsters talk about evangelicals, they usually mean white evangelicals. However, the doctrinal and social marks of evangelicalism likewise apply to Black evangelicals. The difference is political. When white evangelicals seek political allies, they overwhelmingly look to the Republican Party. Most Black evangelicals vote Democrat.[6]

Being an evangelical is different from being Catholic or Baptist or Mormon. These designations are *ways* of practicing a religion. Evangelicalism is a mindset that transcends denominations. Over the last decade, the percentage of Orthodox Christians and Mormons who identify as evangelical has doubled, and the percentage of Catholics who identify as evangelical is steadily increasing.[7]

In fact, identifying as evangelical may have nothing to do with where or even *if* you go to church. A 2016 Pew study looked at white non-evangelical adults who expressed positive views of Donald Trump. By 2020, 16 percent of this same group began describing themselves as evangelical.[8] Fifty percent of evangelicals who say they never go to church are politically conservative.[9] Unchurched but self-identified evangelicals remain theologically and socially conservative, maybe even more so than their churched brethren.[10]

It is not a surprise that evangelicals as a demographic are most closely associated with conservative politics, and specifically, the Republican Party.

That is not by accident.

In 1979, Jerry Falwell created the Moral Majority with the goal of uniting conservative voters, not necessarily evangelicals. The Moral Majority was fueled by religious leaders like Falwell, but also by conservative politicians who wanted to move the Republican Party to the right. It was Falwell's goal to move the country back to the value system, under God, on which America was founded, as he saw it.

Ten years later, evangelicals and the Republican Party formally joined forces.

It was a chance meeting. Ralph Reed, a young Republican political operative, was randomly seated by Pat Robertson at a conservative conference. Robertson saw talent in the young man, and he asked him to lead a new religious right organization.

Here, Reed recounts Pat Robertson's vision for the country: "Operational control of at least one of the two major political parties. Elect a committed Christian as president of the United States, take a majority in Congress and the US Senate, elect a thousand committed Christians—devout Christians at every level of government. This isn't just going to be some Christian civic group. This is going to be the most effective public-policy organization in the country. And at the end of 10 years, American politics is going to look totally different."[11]

Reed went to work. He shaped the new organization, the Christian Coalition, into a daunting pro-traditional family, evangelical political force. Reed identified two hundred thousand evangelical churches in the country, gathered church directories, names, and phone numbers. He sent organizers to the homes of evangelicals who were not registered to vote.

To those evangelicals resistant to the idea of a marriage between faith and party, Reed had this message: "If you choose to disengage, it will show up on your doorstep in public policies that will undermine and assault your beliefs, and elected officials who don't share your values."[12]

In 2016, it was Ralph Reed who was called upon to calm evangelical fears about Donald Trump after the release of the *Access Hollywood* video. Donald Trump isn't your Sunday school teacher, was Reed's message, but he's the president evangelicals need.

Reed, now with the Faith and Freedom Coalition, is focused on bringing non-white groups into the political fold under the banner of social and cultural conservatism.[13]

We cannot talk about evangelical science denial without consideration of conservative politics, whatever the flavor—Republican, Tea Party, or Libertarian. Except for the issue of biblical inerrancy, the social and political inclinations of being evangelical have more influence on evangelical attitudes toward science than do doctrinal beliefs. In many ways, views on science are indistinguishable—is this a political view or a religious conviction?

In the decades following the direct targeting of evangelicals for political purposes, scientists identified more and more health and environmental consequences of technology and industrialization. Conservatives grew disenchanted with proposed government regulations to remedy them. Environmental concerns weren't the bad guy; scientists were.[14]

Republican confidence in science has dropped nearly thirty points since 1975.[15] Predictably, evangelical Christians have among the lowest trust in science and the highest rates of vaccine hesitancy.

Evolution Started It All

Not since Galileo was brought before the Church Inquisition for his sun-centered model of the solar system has science caused so much angst in religious circles.

Of course, there were challenges to evolution from the time Darwin published, but evolution did not enter the collective consciousness of evangelicals until the twentieth century. The Scopes case put evangelical science denial on trial, and evangelicals have been fighting evolution from that day forward.

You may not accept evolution, or you may accept evolution, or you may be some gradation in between. Acceptance, or not, doesn't change the fact that modern biology relies on the theory of evolution to explain everything we know in biology.

The purpose of this book is not to defend or explain the theory of evolution.[16] This book starts with the reality of evolution science: evolution is the foundational principle of all modern biology. Evolution informs modern medicine, modern genetics, modern agriculture, and modern conservation of endangered species.

You may not accept evolution, or you may accept evolution, or you may be some gradation in between. Acceptance, or not, doesn't change the fact that modern biology relies on the theory of evolution to explain everything we know in biology. Acceptance, or not, doesn't change the fact that evolution theory continues to be a powerful predictor of new knowledge.

The purpose of this book is to explore evangelical science denial, and that conversation begins with evolution. How evangelicals talked (and still talk) about the science of evolution shapes the way evangelicals talk about other areas of science.

Creationism

There are many people of faith who accept evolution and an ancient universe and who also acknowledge God as the originator, sustainer, and wisdom of the universe—the Creator of it all. Some even call themselves "evolutionary creationists."

The terms *creationist* and *creationism* as used in this book, however, refer to a specific viewpoint regarding the origins of the universe.

As with any attempt to categorize, there are nuances and gradations. Generally, all creationists reject evolution, reject common ancestry, and believe in special creation.

Creationists are *young-earth creationists*, believing that everything was instantaneously created less than ten thousand years ago, or *old-earth creationists*, allowing (in various ways) for instantaneous creation, but over long periods of time. "Intelligent design" is not a belief separate from creationism, but rather is a way of defending such views.

There are scores of organizations birthed out of opposition to evolution. Although born out of evolution denial, these organizations also promote opposition to climate science and COVID science, as well as opposition to other areas of modern science.

Here are some of the major players:

- Answers in Genesis
- Institute for Creation Research
- Discovery Institute
- Reasons to Believe
- Compass International
- Does God Exist?
- Creation Ministries International
- Creation Today
- Apologetics Press
- Is Genesis History?

These organizations produce far-reaching radio, television, and web-based programming. For example, Answers in Genesis programming is heard on over seven hundred radio stations across America and on hundreds of stations around the world in Latin America, Africa, and Asia, in fifteen different countries.[17]

Other evangelical organizations such as the Gospel Coalition, Grace to You, Desiring God, and Focus on the Family have a broader evangelical scope, but content addressing evolution, climate science, and COVID science is included under their larger banners.

Science on My Side

Science writer Faye Flam launched a podcast during the COVID pandemic.[18] Her goal was to wade through torrents of information and misinformation regarding all things COVID. She chose, or so she thought, a catchy name for her podcast, a name she thought defined her goal as a science communicator. A name free from partisan entanglements.

She regretted her choice almost immediately.

Flam called her podcast *Follow the Science*. She cringes whenever she hears the phrase used as a snarky rhetorical weapon lobbed by opposing sides.

One side lobs the phrase at people wearing N95 masks alone in a car. The other side lobs it at COVID-precaution resisters. "Follow the science" quickly became political.

Senator Rand Paul (R-KY) pushed early and hard for schools to reopen, tweeting "I wonder where the 'listen to the science' people will go when the science doesn't support their fearmongering or their chosen narrative?"[19]

Rarely is the phrase "follow the science" actually followed with science. Often it is launched as an accusation—are you a fearmonger choosing a narrative of fear? Follow the science! Sometimes it is a barrier erected against unsettling ideas. Sometimes it is a defense when long-held religious beliefs are challenged. Sometimes it is wielded against a perceived assault on liberty. Usually, it has nothing to do with science.

We intuitively understand the authority of science. We want the standing science commands. We want science on our side. Even in the case of conflicting positions, all sides claim the authority of science. In the crossfire of conflicting positions, how do we sort it all out?

When the science of evolution entered the consciousness of rank-and-file evangelicals in the 1920s, it shook their world. Theologians and academics had been discussing it, sure, but it wasn't a hot topic at the church picnic.

Then evolution came to the local public school, and it was personal.

And when rank-and-file evangelicals wanted to sort it all out, they did not turn to scientists. Those scientists, after all, want to tell our children that they came from monkeys.

Instead, evangelicals turned to a larger-than-life folk hero. William Jennings Bryan, winning attorney in the Scopes trial, was known as the "Great Commoner" long before the trial. Bryan defended the Bible in a courthouse in little Dayton, but the stage was worldwide. The "Great Commoner" spoke our language. He told us evolution is unscientific, bad for our morals, and un-American, so we went with that.

When modern evangelicals find themselves in the crossfire of conflicting positions about evolution, the age of the earth, climate science, pandemic science, and modern biotechnology, most often, they are not turning to scientists. Most often, they are listening to their pastor. Their Sunday school teacher. A folksy politician. A political pundit who visits their home every night on cable news. All claim to follow the science.

And when they speak against any field of science in the modern era, they still use Bryan's outline: danger to families, to faith, to freedoms.

Whom do we trust? Whom do we believe? It can be overwhelming. Why does the "science" we are supposed to follow change? Are scientists flip-flopping? How do we discern real science from pseudoscience? Are all opinions equally valuable?

The COVID pandemic revealed a deeply ingrained and carefully cultivated distrust of science by those in the evangelical world.

How did we get here?

3

Whom Do You Trust?

You've likely heard of the travels of Flat Stanley, but have you heard of Flatrick Burke?

I first saw him driving around the university campus where I teach. He's hard to miss. To call his silver hatchback a clunker is a compliment, but it's the decor that catches your eye. It's covered in Hawaiian leis and bumper stickers and full-size American flags flying from the windows.

It's driven about town by a well-known local, but this was my first sighting. There were messages on the back window, on the hood, and along the sides in spray paint and stick-on letters: "Scientifically Earth is Flat," "Google Flat Earth," and "Sea level is 100% Level." Burke, whose actual first name is Patrick, was given the nickname "Flatrick" by locals, for obvious reasons.

Patrick Burke is also an evolution and climate-change denier (shocker), but a flat earth is his primary bandwagon of choice. His house, just a half-mile from campus, continues the theme: "Gravity is not real. The earth is motionless." He's done his own research, of course.

The newest addition to his vehicle is painted across the front windshield: "COVID = HOAX." Again, color me shocked.

In the summer of 2020, Burke was issued a citation for criminal trespass after defacing the local Walmart. His graffiti? "COVID is a hoax."

Burke is a college graduate, employed, and a self-described "regular guy." Walmart graffiti notwithstanding, Patrick is not really harming anyone. No one is going to force him to buy a globe.

Now, imagine Patrick or one of his fellow flat-earthers teaching world geography at the local middle school. If he insists on teaching a flat earth, he will be invited to take his teaching skills elsewhere.

In the summer of 2021, employees of the Houston Methodist Hospital system marched carrying protest signs blazoned with "No Forced Vaccines" and "Stop Medical Tyranny." The protest was in response to the suspension, without pay, of 178 workers for refusing to be vaccinated against COVID.

One hundred seventeen of the suspended employees subsequently filed a lawsuit in federal court.

The lawsuit was the first of its kind. It is important to note that the hospital system exempted hundreds of employees from the vaccine requirement for religious or medical reasons and for pregnancy.

US District Judge Lynn Hughes tossed out the lawsuit and made this statement, referring to the lead plaintiff: "Bridges can freely choose to accept or refuse a COVID-19 vaccine; however, if she refuses, she will simply need to work somewhere else."[1]

In the Houston Methodist system, patient safety is a priority. If it is not the priority of an employee, the employee is invited to take their skills elsewhere.

Merit versus Source

How do we decide what is true? How do you decide whom to trust? Which people do you believe? What institutions do you trust? How often is our trust based on direct knowledge?

Here's an easy one—most of us cannot explain the intricacies of aerodynamics. Yet, we trust those who can—engineers, mathematicians, mechanics, pilots—and we get on the plane.

Here's another—most of us are not biochemists or molecular biologists, but, if need be, we will take an over-the-counter pain reliever approved by the Food and Drug Administration (FDA). If we have a bone set or a surgical procedure, we are essentially trusting the work done by the National Institutes of Health (NIH) and other research facilities.

None of us, including scientists, understands the intricacies of all medicine, all technology, all possible fields of biology, chemistry, and physics. We trust experts who have been tasked with understanding the science for us and charged with helping us make informed decisions.

When trying to understand science, do we judge an idea on the merit of the idea or the source of the idea?

In a pandemic due to a respiratory virus spread by coughing, breathing, and sneezing, who would have predicted that a mask would be a bridge too far?

When the city of Stillwater, Oklahoma, mandated masks in public spaces, things got ugly. In the span of three hours after the mandate went into effect, store employees were verbally abused and threatened with physical violence, even with gun violence. The city manager was forced to repeal the mandate for the safety of store owners and employees.[2]

Early in the pandemic, conservative commentators such as Rush Limbaugh, Dennis Prager, and Laura Ingraham fired up their groupies by comparing themselves to Rosa Parks and other civil rights dissidents. Early on, the mask message from conservative media was not about contagion or public health or stopping the spread, but about fear and control.[3]

Instead of a tool in a pandemic, masks became a symbol—a symbol of control by elitist progressives. Masks were called "muzzles" and "face diapers" by conservatives, including evangelical Christians.

Rejection of masks did not arise from a careful, or even cursory, reading of peer-reviewed studies.[4] It was a group decision, and we went with our group.

Authority or Expertise?

Polymerase chain reaction (PCR) is a method used to amplify a tiny bit of genetic material. PCR is like molecular photocopying. Scientists can take a minute sample of DNA or RNA and make billions of copies.

If you experienced a foot-long Q-tip tunneled through your nasal passages as part of a COVID test, you are personally familiar with the RNA-collection step in a PCR.

American biochemist Kary Mullis won the 1993 Nobel Prize in Chemistry for his work with the PCR method. Did Dr. Mullis then rest on his laurels and retire in scientific glory?

No, he did not.

Mullis succumbed to Nobel disease. *Nobel disease* refers to the few scientists who, after making a remarkable discovery, advocate for decidedly *un*scientific, even whacky ideas later in life. After his Nobel, Mullis turned his attention to AIDS. "It's all a hoax!" insisted Mullis. According to Mullis, AIDS is a conspiracy perpetuated by environmentalists, government agencies, and scientists to protect their jobs and incomes.[5]

Mindful of the cautionary tales of Nobel disease, how can we know whom to trust? Apparently, a collection of letters following a name isn't always sufficient.

Back in the day, Mr. Wizard was the only celebrity scientist. Today, people quote Dr. Oz and Dr. Phil far more than their actual family doctors. And during the pandemic, Christian programming featured a succession of doctors pontificating about all things COVID.

Look no further than the steps of the Supreme Court of the United States. A group of doctors calling themselves "America's Frontline Doctors" found more than their fifteen minutes of pandemic fame on evangelical programming. Simone Gold, spokesperson for the group, was an especially popular guest.

We will talk in depth about America's Frontline Doctors in the next chapter, but for now, consider the optics. All in the group have medical training. All have letters following their names and the title of doctor. All are wearing the international symbol of medical respect—a starched white coat. And where are they standing? On the steps of the foremost symbol of wisdom in our nation—the Supreme Court.

All the optics are in place. The group appears to be qualified. They appear to have the authority to speak about All Things COVID.

When someone presents as an authority in a scientific field, ask yourself, Are they appealing to authority because of their qualifications, or are they appealing to authority because of the evidence for their argument? Qualifications are not to be discounted, of course, but is the person qualified in the scientific field to which they are speaking? And, importantly, is the argument they are making backed by the consensus of scientists in that field? What is the collective position of experts based on the body of evidence? The consensus position is not the final word in the matter, but it is definitely the starting point.

You can always find a person with letters after their name who will tell you what you want to hear. It is tempting to cherry-pick an authority with an appealing minority position. But if you ignore the consensus of experts, you are likely judging an idea on the *source* of the idea, rather than on its merits.

It's wise to hold a cautious approach to authority and a robust respect for expertise.

My Own Research

Social media is the twenty-first century's village square, and anyone can be the town crier. As the COVID pandemic descended upon us, social media went into overdrive. Before long, common themes emerged: Doctors are not allowed to treat people with "proven and inexpensive" remedies like ivermectin and hydroxychloroquine, and conversely, doctors and hospitals "refuse" to prescribe alternative treatments. People who take alternative remedies "always get better."

And invariably, the documentation for all such findings is "first-hand accounts."

West Hanson is a social worker in Southeast Texas, a land of oil refineries and Bible churches. Hanson regularly makes home visits to check on the health of his clients.

Most of his clients are older, and almost all of them are resistant to receiving a COVID vaccine.

Hanson's clients are living, breathing proof texts for polling that says white evangelical Christians are the largest demographic resisting vaccination.

Hanson has heard it all. The vaccine is a depopulation tool. The vaccine makes you transhuman. The vaccine connects you biologically to the internet.

Donna and her husband are Hanson's clients, and both are devout Baptists. Donna is a work-from-home professional, and her husband is retired. Neither Donna nor her husband is vaccinated. Curious as to how she arrived at her anti vaccine position, a reporter for National Public Radio asked Donna where she gets her news. Here's Donna: "I don't really watch a news broadcast," she says. "I just do a lot of research, and people that I trust, that feel the same way I do, I follow."[6]

As the COVID virus spread through the country, so did another malady: sudden expertise.

In the land of the free and the home of the independent, "research both sides and make up your own mind" sounds like sensible advice. It sounds American. Suddenly, we no longer need scientists to map out the risks of viral spread. We no longer need double-blind, random, controlled studies to tell us which treatments work and which don't. We certainly do not need the consensus of experts in the fields of infectious disease and epidemiology.

As the COVID virus spread through the country, so did another malady: sudden expertise.

"Everyone has their own facts—you have yours and I have mine," as I was told.

And when the "agree to disagree" card is played, the conversation is done. Tom Nichols (*The Death of Expertise*) calls "agree to disagree" a "conversational fire extinguisher."[7] Insisting that not everything is a matter of opinion does not win friends and influence people.

In a country where everyone's vote counts the same, anyone can be an expert. As long as I "research both sides," my conclusions are as good as yours, or those of any scientist for that matter.

What does personal research look like? Firsthand accounts from those who "think like me"? Do we gravitate toward sources that mirror the prevailing opinion of our group—religious, cultural, or political? Do we pass all subsequent information through the filter of our initial opinion?

Without a doubt, we can find websites and blogs affirming ideas outside the consensus of a scientific field. We can always find people with advanced degrees championing outlier opinions. And in the end, it may feel as if we've done the hard work of wading through the weeds and digging up the real truth.

During the COVID pandemic, to reject the consensus of scientific experts was to be autonomous. I did my own research. I am not one of the sheeple.

Teach the Controversy

How did "research both sides and make up your own mind" become equally applicable to buying a new refrigerator and judging the efficacy of a vaccine in a deadly pandemic? When did science become an exercise in beliefs and values instead of evidence?

Look no further than your local middle school and high school.

Over the last six decades, the teaching of evolution in public schools has been challenged in courts. From outright bans to demands of equal time for creationism, evangelicals fought a battle against evolution science. Although creationism lost in every major federal case, the battle continues at the school board level.

Despite court rulings, a survey found that 13 percent of high school biology teachers explicitly teach creationism at some point in the school year. Although they did not explicitly teach creationism in class, another 29 percent of biology teachers report feeling nervous about evolution at open house or other meetings with parents.[8]

When did science become an exercise in beliefs and values instead of evidence?

It's the nervous teachers (and those who did not admit their reticence) who are impacting our understanding of how science works. All told, 60 percent of the biology teachers in the study were neither strong advocates for evolution nor enthusiastic supporters of creationism. They've been called the "cautious 60%."[9]

Wanting to avoid trouble and the hot seat of parental complaints, the cautious teachers soft-pedal evolution. Sometimes they skip the human part. Sometimes they teach evolution only as it applies to molecular biology. Sometimes they distance themselves from the topic by telling students it doesn't matter if they believe it; just learn it for the test. Often, they "teach the controversy"—informing students about creationist arguments against evolution.

The lukewarm treatment of evolution in the classroom sends an unmistakable message: when it comes to science, you can decide for yourself. Evidence is optional—believe it, or don't. When everyone decides for themselves, values and beliefs trump facts.

Evolution, like all other sciences, has frontiers. There are things at the edges of our knowledge about which we are still learning. Biologists are debating the mechanisms, patterns, and details of evolution, but the fact of evolution is not in question. As we gather more evidence, ideas at the frontiers of any scientific field adapt, change, and adjust.

What doesn't change are the foundations of a field of science. The foundations of science are the theories—atomic theory, gravity theory, germ theory, cell theory, evolution theory. The frontiers are in flux. The theories are not.

When everyone decides for themselves, values and beliefs trump facts.

When legitimate areas of discussion at the frontiers of science are exaggerated, we open the door to alternative explanations for the foundations of science. We cry "teach the controversy" when there is, in fact, no controversy. If the experts in the field are not debating it, it's not a debate.

Teaching the "controversy" of evolution opened the door to a choose-your-own-adventure approach to all science.

It's not a surprise, then, that evangelicals who deny evolution also doubt evidence of climate change and the science of a pandemic. American evangelicals are fostered in free speech and fairness and equality, and everybody gets a say. We assume facts play by our rules.

Just the Facts

Science is a way of knowing. Specifically, science is how we know about the physical world. Science is about evidence—factual, empirical, repeatable, testable evidence. Facts are not subject to our opinions. We can have an opinion as to how we should proceed given a set of facts, but we don't get to have our own facts.

We can have an opinion that the earth is flat, but our opinion doesn't change the fact that the earth is a sphere (more or less).

Science asks: What is knowable? Science looks for natural explanations for natural phenomena. Science is an objective way of knowing. How is it built? How does it behave? How does it work?

Science is a neutral tool.

What happens when scientific evidence runs counter to our religious or political convictions? Emotions kick in. We are uncomfortable. Our impulse is to dismiss the source of discomfort and seek shelter within a like-minded group.

When surrounded by those who are likewise uncomfortable with an aspect of scientific evidence, we are empowered. When surrounded by those who think like we do, it's easy to believe we are on the right side of the issue.

In legitimate scientific inquiry, there is no room for personal biases. Scientific evidence must be reproducible by other scientists who have no motivation to confirm the original findings. Peer review is key to scientific inquiry—more on that in chapter 4. At the frontiers of our scientific knowledge, scientists may quibble, but in the end, the process prevails.

The process makes no room for biases, but that does not mean individual scientists do not have personal biases. Scientists aren't Mr. Spock clones.

Legitimate science, therefore, is done in community. Hypotheses succeed or fail depending on repeatability by others in the scientific field.

When there is legitimate disagreement, scientists take another look. They go back to the natural world, collect more data, and make their case.

But it's still not over. Scientists love to prove each other wrong.

In the end, it's not the opinions of the scientists that matter. The evidence decides the answer to the question at hand. Science tells us when we need to change our minds.

Cooking the Books

Francis Collins, former head of the NIH, was at the helm of the United States' COVID response. Collins, who makes no secret of his Christian faith, was keenly disappointed in his fellow evangelicals and their resistance to COVID vaccines: "I am just basically heartbroken in a circumstance where, as an answer to prayer, vaccines have been developed that turned out to be much better than we dared to hope for."[10] Evangelicals have been "victimized by the misinformation and lies and conspiracies that are floating around, particularly on social media," said Collins.

More than a year after COVID vaccines were released, popular conservative pundit Tucker Carlson characterized public health leaders like Collins as morally bankrupt and complete failures.[11]

Carlson's hot take is not new—evangelicals have a long history of disparaging scientists: scientists mislead, falsify evidence, and fabricate data. Scientists cover up evidence for a young earth and special creation. Scientists lie and cook the books.

Rejection of science by evangelicals started in earnest following the Scopes trial and continues unabated.

Brian Thomas holds a PhD and is identified as a "research scientist" at the Institute for Creation Research (ICR). Thomas warns evangelicals against taking mainstream narratives regarding climate science and human evolution as truth because "equally intelligent and educated people" disagree.[12]

According to Thomas, "conventional thinkers" (read: those who accept the consensus of the scientific community) assume that those who diverge from the mainstream simply need more science education.

Not so, says Thomas: "We've all heard what the secular scientists are saying, and larger doses of it likely won't change our minds."[13]

Information didn't change minds about evolution or climate science, and apparently it did not change minds in the pandemic. Despite the repeated mantra of "follow the science," scientists were suspect.

A study out of the University of New Hampshire put "follow the science" under the microscope. During the pandemic, those who trusted scientists the least were the most resistant to COVID precautions. Individuals who believed scientists adjust their data to fit preconceived conclusions were the least likely to fear the COVID virus, the least likely to wear a mask, and the most likely to prioritize restarting the economy over stopping the spread of the virus.[14]

Interestingly, even those who believe that mainstream scientists are cooking the COVID books identify with Team Science. After all, if you have your own facts, you need your own set of experts. Those who took the livestock dewormer ivermectin to prevent or treat COVID were quick to point out that the drug was developed by a Nobel Prize winner.

We want experts on our side, even if we reject the consensus of the scientific community. Despite rejecting scientific evidence, we intuitively understand its importance.[15]

Science Can't

Science is an objective way of knowing about the universe. Science explains what we observe in the physical world, but science doesn't explain everything.

What can science *not* do? Science cannot explain the meaning of life. Science does not guide morality. Science does not answer the "who" and "why" questions about creation.

We often want science to do all the things. We want science to explain all the answers: Where is God in all this? Where is God's love? Where is God's providence? Where is God's care for creation? Where is God's sovereignty and sustaining hand? We want answers, but these questions cannot be answered by science.

Science cannot determine value. A natural explanation for the water cycle does not devalue the gift of rain in a drought. Children are valued as blessings from God, yet everyone acknowledges the nine-month natural embryologic process that produces a child.

A natural explanation for creation does not demean the value of creation.

Science cannot consider all possible explanations. This aspect of science is disconcerting to many evangelicals. Science only considers what can be observed and tested. We can talk about an intelligence underlying creation, but that is a theological conversation, not a question for science.

"It's only a theory" is a common indictment of evolution. School districts have gone to court for the right to place "only a theory" stickers in biology textbooks. Not only is this a misunderstanding of the term *theory*, but the phrase implies that evolution is simply one idea among many, all worthy of consideration.

Science cannot provide absolute proof. For years, evidence accumulated linking smoking to lung cancer. And for years, tobacco companies shrugged off the mountains of evidence, demanding absolute proof.

We've been demanding absolute proof for evolution since the Scopes trial. Despite a wealth of evidence from the fossil record and modern genetics, creationists claim there is not "a single example of one creature kind morphing into a separate kind."[16]

Ken Ham of Answers in Genesis is well known for a snappy comeback he teaches children to say whenever their teacher talks about evolution.

Teacher: "Here's fossil evidence for evolution."

Ham: "Were you there?"

Teacher: "Here's evidence from geology that the earth is billions of years old."

Ham: "Were you there?"

Teacher: "Here's a fossil of a fish with a wrist and a neck and tetrapod limb bones and it's found in the rock layer between fish and the first four-limbed animals . . ."

Ham: Yawn . . . "But were you there?"

You get the picture.

The absolute proof of an eyewitness account is the only evidence accepted.

Carefully conditioned, we demand absolute proof that a vaccine will have no side effects, no matter how minimal. We demand absolute proof that a vaccine doesn't alter DNA, no matter how impossible. The "Were you there?" equivalent for a COVID vaccine is "Do you know FOR SURE?"

And of course, we demand absolute proof that the climate is changing. We want to see it happen, in real time, right before our eyes.

Status Reports

Science is never done. We can always learn more. Science is both reliable and subject to change, at the same time. We look at the best evidence at hand. Sometimes we find out we were wrong.

She's not a scientist, nor does she play one on TV, but *Parks and Recreation*'s Leslie Knope knows a thing or two about how science works. When asked why she flip-flopped on an issue, Knope responds: "Well, because I learned new information. When I was four I thought that chocolate milk came from brown cows, and then I flip-flopped when I found out there was something called chocolate syrup."[17]

Very early in the COVID pandemic, the Centers for Disease Control and Prevention (CDC) suspected that asymptomatic people could spread the virus. But this was early in the whole ordeal and more data were needed. By October 2020, with more data, the CDC was confident that asymptomatic airborne spread was occurring.

In March 2020, months *before* the CDC was confident about asymptomatic spread, Anthony Fauci, director of the National Institute of Allergy and Infectious Disease, spoke about masking: "The masks are important for someone who's infected to prevent them from infecting someone else. . . . Right now in the United States, people should not be walking around with masks. . . . When you think masks, you should think of health-care providers needing them and people who are ill."[18]

There are three key points in Fauci's statement: (1) Infected people should mask. (2) *Right now*, asymptomatic people don't need to walk around with a mask. (3) Health-care providers are the masking priority.

When the CDC was certain that asymptomatic people were spreading the virus, masks made sense for all of us. With more information, we changed course.

Oh, but Anthony Fauci's March interview went viral. Selectively edited clips of Fauci saying "people should not be walking around with masks" flooded social media. Flip-flopping Fauci was the meme du jour. Get the tar and bring the feathers.

In the same March interview, Fauci went on to say that masking for everyone could lead to a shortage of masks for those who really need them. At this point, hospitals were slammed with COVID patients and production for masks had not yet ramped up.

Also at this point, Americans were hoarding paper towels and toilet paper. "Leave some for the next guy" was not a part of our early pandemic vocabulary. When we weren't sure about asymptomatic spread, we saved the few masks we had for medical personnel.

Science is both reliable and subject to change. Science lets us know when we need to change our minds. When new evidence means a change of course, that's not flip-flopping. That's smart.

Science writer Carl Zimmer put it best—a scientific paper is never a revelation of absolute truth. At best, it is a status report.[19]

How do we decide whom to trust? How do we decide what is true?

4

Scientific Literacy in a Time of COVID

It was a typical birthday party for a ten-year-old boy—too much sugar, balloons, rambunctious kids. Oh, and party favors, too!

But this was not your usual grab bag of gum and cheap toys. The loot bags also contained a British five-pound note, worth about $10.00 at the time. There was one small catch. . . .

In order to get the party prize, each little guest had to let the birthday boy's dad draw a blood sample.

Chuckling, Dad later recounted the hilariousness of the situation. Line up the kids and they stick out their arms. The birthday boy's best friend faints. One kid throws up all over his mother. Just your standard ten-year-old's birthday party. A laugh a minute.

And you thought bounce houses got out of control.

Dr. Andrew Wakefield used the blood samples he collected at his son's birthday party in one of the most notorious medical studies in modern times. Wakefield's study, published in the prestigious British medical journal *Lancet*, triggered a decades-long collapse of public confidence in one of the most significant medical advances in human history: vaccination.

In his 1998 article, Wakefield claimed that in eight children, the onset of autism followed immunization with the measles-mumps-rubella (MMR) vaccine. Researchers all over the world tried to replicate his results, but no one could. Meanwhile, vaccination rates plummeted in the UK and in the United States.

The thing is . . . Wakefield made it all up.

The birthday party was just the beginning. In addition to an uncontrolled study with sketchy "volunteer" recruitment, he fabricated data. Before the *Lancet* article was published, Wakefield filed a patent for his own version of the MMR vaccine. Wakefield was also being paid for his expert testimony in a lawsuit against vaccine manufacturers.[1]

Twelve years after publication, the *Lancet* retracted the article and Wakefield lost his medical license. Decades of worldwide research and tens of thousands of cases demonstrate no link between autism and vaccination.

Unfortunately, it has been much harder to convince the general public.

In 2015, the largest measles outbreak in decades was traced to Disneyland, which instead of being the happiest place on earth was apparently the spottiest place. The primary reason for the outbreak? Unvaccinated guests.[2]

In 2008, the Somali immigrant community in Minnesota was concerned about autism in their children. The autism rate for Somali children was no higher than that of other populations of children, but once a story becomes *the* story, perception becomes fact.

Leaders of the community invited Andrew Wakefield to speak to parents, despite the loss of his license. In 2008, the vaccination rate of Somali immigrant children matched the general population, about 92 percent. By 2016, the vaccination rate was just 42 percent. In 2017, there were more measles cases in Minnesota than in the entire United States.[3]

Unrepentant, Wakefield continues to travel and speak to anti-vaccine groups. Despite the loss of a measles-free status in the UK and multiple measles outbreaks in the United States, groups continue to pay Wakefield thousands of dollars for his "expertise." To celebrities and everyday moms and dads who do not want to "poison" their children, Wakefield is a martyr.

Anti-vaccination attitudes are nuanced and varied. But in this case, it started with bad science and is sustained by a lack of scientific literacy.

Control

It was only the size of a beach ball, but it produced a tidal wave. In 1957, the Soviets put the first satellite into earth orbit. Like a starting pistol in the race to space, the launch of Sputnik sent Americans scurrying to catch up. Catching up meant the National Defense Education Act, new textbooks, new programs, and over a billion dollars poured into science education.

In the wake of Sputnik, we measured learning by giving tests. Scientific literacy meant facts. The more facts you knew, the more scientifically literate you were. So, we memorized—the planets (in order, of course), the periodic table (just long enough for the test), and the visible light spectrum (Roy G. Biv). We equated knowledge of scientific facts with scientific literacy. We knew facts, but not how science works.

You've heard of the scientific method. At some point in your schooling, you've probably memorized the steps of the method in a particular order. You have my permission to forget that drill. The scientific method is not a lockstep, one-after-the-other, always-the-same recipe. Seldom do scientists work that way. Scientists back up, rework, do more observations, tweak a hypothesis, refine a test, go back, test again. Backtracking and rethinking are part of the process.

There are, however, some nonnegotiables.

Control is one of them. Science experiments use the term *control* in two ways: a control group and controlling variables.

In a scientific study, there are typically two kinds of groups. One group gets the "treatment" (whatever that is), and the other group does not. In a strong experimental design, the individuals in these two groups are as alike as possible.

In an education experiment, the treatment might be a new method of teaching multiplication to third graders. The experimental group of students is taught using the new method, while the control group of students is not. However, we want both groups to be as alike as possible—all third graders, for starters. The more alike the groups are, the more confident we are that the outcome is due to our treatment.

For example, we can't teach a group of third graders with our new method and compare them to a control group of three-year-olds and say "Yay! The new method works!" Likewise, we can't use three-year-olds for our experimental group and third graders for our control group and conclude "This method doesn't work!"

But try as you might to make your control group and your treatment group alike, people vary. A good experimental design takes these variables into account.

If you are designing an experiment for a new Alzheimer's dementia treatment, you will probably include only patients without other health conditions. Here's why: If the patient does not improve after the treatment, is it because (1) the treatment doesn't work or (2) the patient has an underlying condition interfering with the treatment?

Before the first COVID-19 vaccines launched, I signed up for a vaccine trial. I did not care if I was in the treatment group (vaccine) or the control group (placebo). History was being made, and I wanted to be a part of it.

I passed the first two rounds of interviews for the study. I was sitting in the lab awaiting the third and final interview, to be followed by the first jab. I took a hopeful selfie in anticipation of blogging my vaccine trial experience.

Alas, I failed.

Ironically, the autoimmune disease that booted me from the vaccine trial qualified me for 1b vaccination status. I was very disappointed, but I understood. Researchers need to be sure—are unusual post-vaccine reactions due to the vaccine or to an underlying condition?

A good study is a well-controlled study.

A good study also controls for bias whenever possible. A researcher who knows which patients are receiving a new treatment may treat those patients differently, without even realizing it. A patient who knows they are receiving treatment may feel better just knowing they are getting the latest new thing.

The best studies are double-blind studies. In a double-blind study, neither the researcher nor the subject knows who is getting the treatment and who is getting a placebo.

A good study controls for researcher and subject bias.

Peer Pressure

Have you heard this famous quote by Abraham Lincoln? "The problem with things you read on the internet is that they are often not true." And . . . there's your dad joke for the day.

Peer review is an inspection sticker for scientific research. Trust Mr. Lincoln—just because something's printed doesn't mean it's true.

Peer review is a long, arduous process, and it is not for the faint of heart. A researcher or research team submits their study to a journal to be reviewed by an expert in that field of research. If the editors determine the research is worth a deeper look, the study is sent out to multiple specialists in the field.

At this point, the reviewers whip out their fine-tooth combs. Everything is fair game—the groups, the controls, the methods, the tests, the data gathering, the data, the data analysis—you get the picture. The reviewers can reject it, send it back for more work, or in rare instances, accept the study as is.

A study published in a peer-reviewed journal is still not in the clear. If other scientists using similar methods under similar conditions cannot replicate the findings, we've got a problem.

Science is truth-seeking. Science doesn't seek to win for the sake of winning. Science actively seeks to prove itself wrong.

A scientist begins by observing something in the natural world. The scientist then proposes a possible explanation for what is observed. The possible explanation is called a *hypothesis*.

Next, the scientist designs a test for the hypothesis. This is the experiment.

Here's an important point: After the experiment, scientists don't say things like "Lucky me! I was right about my hypothesis!" (At least, they don't say that in their journal report.) If the hypothesis is wrong, the report reads "We reject the hypothesis." If the hypothesis is correct, the report reads "We failed to reject the hypothesis."

A bit convoluted, but you get the point. Good researchers perform tests that could prove their idea is wrong. Good researchers are transparent about their methods and data so other researchers can do the same.[4]

And there's nothing scientists love more than to prove each other wrong.[5] Think about that next time you hear about a vast conspiracy of scientists colluding to hide the truth about evolution or COVID or climate science.

Peer pressure in science is a good thing.

Soon after the publication of Andrew Wakefield's measles-vaccine-causes-autism study, many critics gave the report the skeptical side-eye.[6] For starters, the number of children in the study was very small—only twelve—and there were no controls. None. Throw the research-design red flag.

The data relied on parent recall and beliefs. And of course, there's that matter of blood samples taken at a birthday party. Red flags for experimental control and ethics.

And when other researchers tried to replicate Wakefield's findings, they could not. Wakefield himself refused to attempt replication. Red flags for experimental results.

All the red flags littering the ground were just the beginning. Wakefield had manipulated and misrepresented his data. And while Wakefield was busy discrediting the MMR vaccine, he was filing patents for an MMR-alternative vaccine.

Andrew Wakefield is a cautionary tale in the importance of peer review.

Peer review initially failed. Wakefield's study made the pages of the prestigious *Lancet* journal, but the second phase of peer review exposed the fraud.

Scientific Literacy on the Steps of the Supreme Court

On a sunny day in Washington, DC, just months after the COVID pandemic first gripped the country, a small group gathered on the steps of the Supreme Court of the United States to hold a press conference.

With our nation's most iconic symbol of justice in the background, the group assembled for photos. Each wore a pristine white coat embroidered with a caduceus—two snakes winding around a winged staff—instantly recognizable as the symbol of a medical doctor.

This caduceus, however, was far different from most. The wings were red, white, and blue, with stars spangling the blue. The snakes were striped red and white. Meet "America's Frontline Doctors."

The press conference, billed as the "White Coat Summit," was live streamed by Breitbart News, and it hit the internet in a firestorm. But almost as soon as they hit the web, America's Frontline Doctors were gone, removed by censors and swept into the internet abyss for "violating community guidelines." All the big guns took it down—Facebook, Twitter, Google.

It was too late. When all you have to do is hit "record" on your device, there's just no stopping the spread. The press conference video popped up like little prairie dogs in internet infinity. President Trump shared the video with his eighty-four million Twitter followers. By evening on the day of its release, the video had racked up twenty million views on Facebook alone.[7]

With evangelistic fervor, the spokesperson for the group, Dr. Simone Gold, preached a message of captivity. "We're America's Frontline Doctors," she said. We are here to help Americans heal and to set them free. But it's not a virus holding us captive, according to Dr. Gold—it's fear:

"We are not held down by the virus as much as we're being held down by the spider web of fear. That spiderweb is all around us and it's constricting us and it's draining the lifeblood of the American people, American society, and American economy."[8]

As various doctors took the mic, they spoke about masks (we don't need them). They spoke about school shutdowns (a ploy by liberal unions to get money). Mostly, they spoke about the cures for COVID that "fake science" and "fake pharma" don't want you to know about.

According to America's Frontline Doctors, zinc, Zithromax, and hydroxychloroquine are simple cures purposely withheld from the American people. These three "cures" were praised, but hydroxychloroquine was the star.

One doctor, Stella Immanuel, claimed to have treated and cured 350 patients with hydroxychloroquine. This claim perked up the ears of at least one attendee at the press conference. "Have you been able to publish your findings and results?" the speaker asked. Dr. Immanuel was indignant: "My data will come out. When that comes out. That's great. But right now people are dying. So my data is not important for you to see patients. I'm saying that to my colleagues out there that talk about data, data, data."[9]

Ok then. You'll get it when you get it.

Dr. Gold interjected: "Not every clinician needs to publish their data to be taken seriously."[10] Your data are not important? You don't need to publish? So . . . are we just supposed to take your word for it?

Earlier in the press conference, Dr. Immanuel volunteered her opinion of another well-established research practice: "And let me tell you something, all you fake doctors out there that tell me, 'Yeah. I want a double blinded study.' I just tell you, quit sounding like a computer, double blinded, double blinded. I don't know whether your chips are malfunctioning, but I'm a real doctor."[11]

Don't come at me with your silly old scientific method.

On the first day of each semester in my introductory biology course, I include a photo of America's Frontline Doctors on the steps of the Supreme Court in my PowerPoint slides. Each semester begins with an introduction to scientific thinking—what makes good science? How do we judge the merits of a scientific claim? I tell the story of America's Frontline Doctors, and without fail, even college freshmen can spot the red flags of bad research.

Miracle Cures

It's a loathsome disease. It causes incessant itching, and sufferers often scratch their skin with a stick or rock until they bleed.

Fast-flowing rivers and streams, often sources of fresh water for remote villages, provide the perfect breeding ground for black flies. The bite of a black fly transmits a larval worm. And as the worms multiply, they burrow under the skin.

As miserable as it is, it is not the itching that gives this horrid disease its name. The worms eventually migrate to the eyes, causing infection and inflammation of the corneas and destroying vision.

River blindness debilitates, but it doesn't kill. River blindness spreads its misery in tropical countries—more than 99 percent of river blindness cases are in Africa.

In the early 1980s, researchers discovered a powerful antiparasitic compound in a dirt sample dug up near a golf course in Tokyo. The compound proved to be a powerful treatment for parasites in livestock, and a version was developed to combat the misery of parasitic infections in humans, including river blindness. In 2015, Satoshi Omura and William Campbell shared the Nobel Prize in Medicine for discovery of this drug, capable of preventing so much human misery. The drug? Ivermectin.

Somewhere amid all the pandemic noise, ivermectin found its moment of fame as a wonder drug for COVID that the medical establishment "doesn't want us to know about." "What are they trying to hide?" they say. "It's a Nobel Prize-winning drug," it's argued.

But instead of doctor-prescribed doses appropriate for humans, ivermectin paste intended for livestock flew off the shelves of feed stores and farm suppliers. Patients with nausea, vomiting, hallucinations, blurred vision, low blood pressure, seizures, and coma flooded emergency rooms. In just one month, poison control centers across the country reported a 245 percent jump in ivermectin exposure.[12]

And even in human-sized doses, ivermectin treats infections caused by parasitic worms and arthropods, not viruses.

So why the hype? Early in the pandemic, a study found that ivermectin inhibits the growth of the severe acute respiratory syndrome coronavirus 2 (SARS-CoV-2) virus in cell cultures—in petri dishes, in a lab. Yay, right? Not so fast.

In order to achieve the dose of ivermectin used in the petri dish experiment in actual humans, the concentration would have to be one hundred times higher than the dose approved for use in humans—a toxic level. Almost any drug in a high enough concentration will kill a pathogen—the dose makes the poison, as they say.

Multiple studies followed: Is ivermectin effective as a preventative? Does it reduce COVID-19 symptoms? Does it reduce death or severe disease? Does it prevent hospitalization?

One of the largest studies, released as a "preprint" prior to publication, claimed an astounding 90 percent reduction in death rates with ivermectin. This almost-too-good-to-be-true claim immediately caught the attention of a small group of researchers.

The study was filled with errors. There were inconsistencies between the raw data and the data reported in the paper. There were duplicated patient records. Some patients listed in the study died before the study began. Some of the numbers reported were too perfect to be realistically believable. And there was a pesky problem of plagiarism, too, but oh well.[13]

The study never made it to a peer-reviewed journal, but peer review did its job. The paper was withdrawn from the research platform where it had been posted. Before it was withdrawn, however, it was viewed more than 150,000 times and cited more than thirty times. And once it hit the internet and went out to the general public, well, you know what Abraham Lincoln said.

Many studies later, the answer is a resounding no. Large, randomized, double-blind studies found ivermectin to be no more effective than a placebo in treating or preventing serious COVID illness.[14]

Before ivermectin was the drug they didn't want you to know about, hydroxychloroquine was center stage. Like ivermectin, hydroxychloroquine derives from a substance found in nature. Actually, it has been around in some form since the seventeenth century. A Jesuit priest serving in South America took advantage of an indigenous cure—the bark of the "fever tree"—when he fell ill.[15] Quinine, as chemists named it, soon became a common ingredient in patent medicines.

During World War I, quinine's anti-malaria properties were discovered. Hydroxychloroquine is a derivative still used today to treat malaria and also autoimmune diseases like lupus and rheumatoid arthritis. Hydroxychloroquine is old, cheap, and no one holds a patent for it.

Hydroxychloroquine had been studied as an antiviral in cells, in a lab, in a petri dish. Would this cheap and common drug be effective against the SARS-CoV-2 virus? When the pandemic broke out and we had few other choices, it was reasonable to try it.

The FDA moved at lightning speed to approve the first large, randomized, controlled, double-blind clinical study—in actual people—of hydroxychloroquine and COVID in March 2020. The World Health Organization (WHO) and the NIH both launched clinical trials. Clinical trials also launched in the UK, with the benefit of the National Health Service's databases.

On the same day as the first US trial was approved to begin, a document—not a research paper—hit the internet. The document claimed that hydroxychloroquine works against COVID—not research, a claim. But it was off to the races.

The document first gained momentum on the West Coast and drew the attention of investors. Venture capitalists were ready to go. A former executive of PayPal, LinkedIn, and Square tweeted, "Randomized controls are horrible ideas."[16] Controlled studies take too dang long, said the investors. Let's just give it to people and get this show on the road.

Hydroxychloroquine now had the bullhorn of the COVID-is-no-big-deal, don't-shut-my-economy-down crowd. Before March was over, President Trump was pitching hydroxychloroquine as an "approved" drug. Approved, yes, for malaria and some autoimmune diseases.

It was a relatively safe drug, yes. But effective against COVID? We did not know.

By June 2020, all major trials in the US and in the UK were halted, but not for safety reasons. Hydroxychloroquine simply didn't work. There was no difference in outcome whether the patient was given hydroxychloroquine or a placebo.[17]

A month later, America's Frontline Doctors were on the steps of the Supreme Court telling the internet that hydroxychloroquine was the cure for COVID.

A year later, Simone Gold, the face of America's Frontline Doctors, spoke to a crowd at an evangelical church in Thousand Oaks, telling them to avoid the COVID vaccines and to stock up on hydroxychloroquine and other drugs not shown to be effective against COVID.[18]

Those We Trust

America's Frontline Doctors were a huge hit with evangelicals. In fact, the group was so overwhelmingly popular with evangelicals that it begat a Facebook group, but probably not the kind of group you'd think.

Heather Mashal, an evangelical Christian from Delaware, created the group "Christians Against Covid Denialism" with the tagline "Do you wear a mask in public? Have you been vaccinated against COVID? Do you consider the infamous 'America's Frontline Doctors' to be quacks?"[19]

Evangelical support of America's Frontline Doctors was not by accident. Billed as a grassroots effort by concerned doctors, the group was actually a creation of the Council for National Policy (CNP), an organization with its origins in the conservative resurgence in the Southern Baptist Convention. CNP's organizers, leadership, and supporters include evangelical heavy hitters like James Dobson, Tim LaHaye, Tony Perkins, Ralph Reed, and Paul Weyrich.[20]

Following the press conference on the steps of the Supreme Court, Simone Gold spoke in churches and made the rounds across evangelical media. Popular homeschool speaker and podcaster Heidi St. John could hardly contain her enthusiasm during an interview with Gold, gushing over her new "best friend."[21]

Gold was also interviewed on Daystar Television's Ministry Today show, along with notorious anti-vaxxer Robert F. Kennedy Jr. Host Marcus Lamb ended Gold's segment by telling the audience that Gold was fired from her hospital job for doing what the Bible says to do—"to speak the truth in love."[22]

The Daystar Television Network, one of the two largest Christian television networks in the world, not only featured Simone Gold and America's Frontline Doctors, but they also regularly promoted ivermectin, vitamins, and supplements as cures and preventatives. Network head Marcus Lamb called ivermectin a "miracle drug."[23]

Evangelicals were listening.

Team Sports

The days of reading an opposing viewpoint in an actual newspaper and absorbing information independently, says sociologist Zeynep Tufekci, are over. With the advent of social media, information exchange is a team sport. When we read contrary viewpoints, it's like "hearing them from the opposing team while sitting with our fellow fans in a football stadium. Online, we're connected with our communities, and we seek approval from our like-minded peers. We bond with our team by yelling at the fans of the other one."[24]

We listen to those we trust, even if those we trust stand against the consensus of the scientific community. We are drawn to those who think like we do.

We've been doing it for decades with evolution: It just doesn't make sense! I don't understand how a fish turns into a monkey. Where are all the missing links? How can random evolution produce such complexity and beauty? I didn't come from a monkey—I am made in God's image! I don't understand it, but I believe the Bible. God did it, and that's all I care about. Who needs evidence?

Without an understanding of how science works, we go with what makes sense. What feels right.

Without an understanding of how science works, we go with what makes sense. What feels right. What our heart tells us.

We go with our team.

5

Faith over Fear

She took a call from the loading dock: Your package is here.

Interestingly, the package didn't arrive by plane. This package was placed on a truck and given a special ride from Boston to Bethesda.

Can you bring it up? she asked.

No, they said. You must come downstairs and meet the driver. And bring your ID. We can only give the package to you.

I imagine she ran all the way.

She was young, just thirty-five, and thoroughly a member of the selfie generation. She asked the driver to take a photo of her with the box.

And he was like, No, ma'am, that's not my job.

Elated, she took the box back to her lab, where 250 little mice awaited. The box contained doses of a COVID-19 vaccine, developed using her scientific research. Each little mousie was about to get a jab.

And the young woman was about to save the world.

Meet Dr. Kizzmekia Corbett. She's young and she's brilliant, and she was the lead researcher in the NIH Vaccine Research Center's lab for development of coronavirus vaccines. She was the primary scientist behind Moderna's COVID vaccine.

Unbelievably, in just a few months, a scientific concept in Dr. Corbett's laboratory became a nationally distributed vaccine that is 94 percent effective against serious disease. It was far from beginner's luck.

Corbett had been studying coronaviruses for more than six years when the COVID pandemic struck. Her attention was on vaccines for Middle East respiratory syndrome (MERS) and severe acute respiratory syndrome (SARS)—coronaviruses that put the world on the edge of a pandemic but stopped just short.

On December 31, 2019, a respiratory illness caused by a coronavirus was reported in China. Emails to Corbett from Anthony Fauci and Barney Graham (Corbett's boss at the Vaccine Research Center) arrived in January. "Buckle up," they told her.

By January 10, 2020, researchers published the genetic sequence of the virus that causes COVID-19.

Sixty-six days later, a vaccine developed in Corbett's lab entered the first phase of human trials. That speedy timeline made some people really nervous.

When questioned about the speed of the vaccine from lab to arms, Dr. Corbett gave a surprising answer: It could have been faster. "We didn't quite get there for a MERS or SARS vaccine," she said, "but that research got us ready for COVID-19."[1]

Kizzmekia Corbett is a science rock star. And Kizzmekia Corbett is a scientist of deep faith. She is a Christian who makes no secret of her love for Jesus.

Corbett slept very little during the pandemic, and she worked seven days a week. On Sundays, she stopped to watch a recorded church service and then spent the remainder of the day sorting through COVID data.

Corbett feels a deep sense of obligation to community health. She sees her work in vaccine development as a way to love her neighbor as herself.[2]

Here's Dr. Corbett: "My religion tells me why I should want to help people, make the world a better place. Science shows me how to study the coronavirus and do the work that one day, hopefully, will prevent people from dying of COVID-19."[3]

Kizzmekia was the kid who entered and won all the school science fairs. When the Nobel Prizes were announced, she wrote speeches and delivered them out loud, with pomp and spectacle and dramatic tears. Francis Collins, director of the NIH and where the buck stopped regarding all things pandemic, said that Dr. Corbett and Dr. Graham were in discussions for "prizes." Dr. Corbett, I hope you kept those speeches.

Who's Missing?

The selfies started hitting social media feeds in December 2020 as critical care nurses, doctors, and hospital staff lined up for the first COVID vaccines. Seeing the photos with tired eyes and smiles hidden behind masks, many of us couldn't help but tear up as the pictures filled our timelines and newsfeeds.

It was historical. It was monumental. It was a victory for researchers and vaccine developers, and a godsend for medical workers.

By February, vaccine availability opened to the first groups of nonmedical people. I received my first dose of the Moderna vaccine at a colossal drive-through at Texas Motor Speedway. Drones and helicopters flew overhead, and the media was about with mics and cameras.

Volunteers and paramedics and medical staff were there by the hundreds, waving and smiling and chatting, and all the while maintaining efficiency like you just can't believe. People were rolling down their car windows and thanking the staff and the volunteers.

I witnessed the same scene at a mega center in Dallas where my eighty-two-year-old mother-in-law received her first jab. The event was electric: big smiles, hopefulness, thankfulness, celebration.

March 2020 seemed like a million years ago.

Before the vaccine launch, we feared that Black Americans would be resistant to vaccination due to a well-justified mistrust of the health-care system. In actuality, the worry was misplaced. There was, however, one large demographic group noticeably absent from the vaccination celebrations: white evangelical Christians.

As vaccines rolled out to the general population in 2021, 45 percent of white evangelicals said they definitely would not or probably would not get a COVID vaccine. Compare that to 90 percent of atheists who said they definitely would or probably would get a COVID vaccine.[4]

More than a year after the vaccine rollout, white evangelicals still lagged behind all other religious groups in COVID vaccination.[5]

As I write, COVID deaths in the United States have topped one million, driven primarily by people who are not vaccinated.[6]

The Pseudoscience of Certainty

For decades, evangelicals have been told that evolution is impossible, just so much made-up foolishness. Henry Morris, the architect of modern young-earth creationism, calls it as he sees it: "Creationists prefer the reasonable faith of creationism, which is supported by all the real scientific evidence, to the credulous faith of evolutionism, which is supported by *no* real scientific evidence."[7]

If scientists have been lying to us for decades about evolution, maybe they are still lying to us. Maybe the seriousness of the virus is overblown. Maybe COVID case numbers are exaggerated. Maybe scientists are hiding cures and forcing their vaccines on us. Maybe *we* are the research!

The COVID conspiracists often told a more compelling story, a more certain story. With tales of microchips and globalists and Big Pharma and Big Science and other big baddies, it was clear who the villains were.

The tales of conspiracies were told with the energy of a sports championship, like an all-out pep rally where we cheered the downfall of our opponents. At the 2021 Conservative Political Action Conference (CPAC), Rep. Lauren Boebert (R-CO) fired up the crowd: "Don't come knocking on my door with your 'Fauci ouchie.'"[8] Later, the crowd erupted in cheers when it was announced that President Biden's 90 percent vaccination goal fell short.[9]

Pseudoscience provides the simplicity and certainty we crave in uncertain times.

Meanwhile, science changes as evidence accumulates, often at a dizzying pace. Science doesn't speak in terms of certainty. Science doesn't speak in terms of proof. Science talks about what the best evidence says.

Without a doubt, there were many failures in science communication during the COVID pandemic. Scientists, politicians, the media, and the public do not speak the same language. Evidence, data, and statistics were updated daily, even hourly. When coupled with a breakdown in general understanding of how science works, the story becomes garbled.

Add decades of mistrust in science by evangelicals to uncertain times, and the stage is set for science denial.

Whistling Past the Graveyard

I'm sure my friend Martin has done many heroic things in his life, but I know for sure he was a hero when he was only eight years old. In 1954, Martin was one of an unprecedented two million children in the largest vaccine trial to that date. He stood in a line, and polio pioneer that he was, put out his eight-year-old arm for the jab. Martin still has his original "Polio Pioneer" button! Martin is a vaccine hero. He personally shares responsibility for saving millions of children from death and paralysis.

Texas epidemiologist Emily Smith was a beacon of information, encouragement, and Christian witness during the COVID pandemic. In an article about kid COVID vaccines, Smith told of one little friend who is scared of needles. Scared as he was, he wrote his best buddy's name on a Post-it Note, took it to the clinic, and repeated his buddy's name over and over while getting the jab.

His buddy is high risk. And this little guy understood that scared though he was, his vaccination would protect his buddy. Faith over fear indeed.

Put *that* on a T-shirt.

In an all-too-literal version of whistling past the graveyard, many Americans put on a brave face (maskless, of course) and announced their fearlessness. "I really don't even give a mask a thought anymore!" one Facebook post proclaimed. "It's irrelevant! I feel sad for the people who are still so scared."

Sen. Rand Paul (R-KY) whistled the same tune: "Will we allow these people to use fear and propaganda to do further harm to our society, economy, and children?"[10] Variations of the meme "fear is the real virus" embellished t-shirts and home-made signs.

For evangelicals, fearlessness became an article of faith. If heaven is the destination of a Christian, why not hasten the transport? If you want to wear a mask, go for it, but as for me and my house, when it's our time to go be with the Lord, we're ready. We don't fear cancer, car accidents, or COVID.

More than simply leaving their fate in the hands of God, faith-over-fear evangelicals wore fearlessness as a badge. Sometimes the badge read "God will protect me from the virus." Sometimes it read "Bring it on."

Joy Pullmann wears the "bring it on" badge. Writing for *The Federalist*, Pullmann recounts historical Christian martyrdom as well as ongoing modern persecutions. Churches who take COVID precautions exemplify the weakness of the modern Western church, according to Pullmann. "God decides when we die," she writes, so why fight it?[11] Pullmann calls on Christians (individually) and the church (corporately) to repent of the sins of masking, social distancing, and Zoom church.

Many "bring it on" evangelicals downplayed the seriousness of the virus. According to one creationist website, "they" lied about the numbers and used the "fear of death" in order to "mass vaccinate people against a flu that has a . . . 99.9% survival rate if properly treated" (meaning hydroxychloroquine and ivermectin, of course).[12]

Marcus Lamb preached fearlessness wearing the badge of "God's protection"—whether that means avoiding COVID altogether or a quick recovery from it, God has you covered.

Marcus Lamb and his wife, Joni, cofounded the massive Daystar Television Network. Daystar reaches more than two billion people worldwide, and is in more than one million homes in the US.

Instead of receiving a vaccine, Lamb told his followers to pray. In addition to prayer, Lamb prescribed ivermectin, hydroxychloroquine, and budesonide (an asthma medication) as foolproof treatments should infection occur. Lamb frequently hosted anti-vaccine activists like Robert Kennedy Jr. and Del Bigtree.

Lamb was not alone in his reliance on God's protection, not by a long shot.

A pastor in Louisiana told his flock not to worry about the virus. He told them to reject the vaccine. He promised that God would protect them. The pastor died of COVID.[13]

The list goes on: A popular Christian musician who disparaged the COVID response as "hysteria" died from COVID.[14] A pastor claimed God told him personally that he would never get the virus, but he did.[15] A Pentecostal denomination lost thirty pastors to COVID, all of whom claimed the protection of God but rejected precautions.[16]

Faith over fear in a pandemic was a top-of-the-temple moment for evangelicals.

Of all the faith-over-fear passages in the Bible quoted in the context of COVID, Matthew 4:5–7 is one passage conspicuously absent. In the ultimate game of "dare ya," Satan took Jesus to the highest point of the temple.

"Jump," said Satan. "Go on, what do you have to fear? Don't you have faith? Throw precaution to the wind! God will protect you."

Faith over fear in a pandemic was a top-of-the-temple moment for evangelicals.

The Real Danger

Nobody claims the COVID virus isn't real, at least nobody who is taken seriously. The real danger, however, does not come from a microscopic piece of nucleic acid wrapped in a spiky protein shell. For many evangelicals, the real danger comes from the players behind the pandemic. The hidden hand pulling the strings is none other than Satan.

Owen Strachan is a theologian, author, and rising star in the evangelical world. When John MacArthur and his Grace Community Church made headlines for their defiance of COVID precautions, there was no bigger cheerleader than Strachan.

The COVID pandemic, according to Strachan, was part of Satan's twenty-year plan.[17] When the neo-Reformed movement took off in the evangelical world in the early years of the twenty-first century, Satan planned a counterattack. Satan would never abide such a resurgence of "sound doctrine." "Wokeness," says Strachan, was Satan's first strike.

The second satanic strike came in the form of a worldwide pandemic, but not for reasons you might think. The evil wrought by Satan in the COVID pandemic was *not* separation, disease, suffering, and death. The real evil wrought by Satan was closed churches.

Here's Strachan: "MacArthur took an equally momentous stand against the global big-government push to close down the church, another evil ministration of Satan that will go down in the history books, and that is part of a far bigger move of a fallen order that hates the worship of Christ."[18] Strachan paused the story to downplay the seriousness of the virus, calling COVID precautions "liberty-destroying, business-ruining, spiritual-health-crashing."

Then, picking up the Satan narrative, Strachan continues: "As events transpired, and Satan's forces moved swiftly to rout the church and drive it from the field, MacArthur (and Grace Community Church elders) saw what was happening."[19]

But Satan wasn't content with simply canceling worship services. Apparently, Satan also has it out for non-FDA-approved COVID treatments.

In November 2021, Marcus Lamb was absent from his Daystar Network show.

He had COVID.

His son, Jonathan Lamb, described his father's infection as a "spiritual attack from the enemy." Satan, according to Jonathan, was displeased because his father championed alternative therapies for COVID: "There's no doubt that the enemy is not happy about that. And he's doing everything he can to take down my dad."[20]

Marcus Lamb died from COVID—but not because he was unvaccinated despite having diabetes, said his wife.[21] Not because of his reliance on the array of treatments he touted on his network, said his son. According to Lamb's family, it was Satan.

Creationist websites took up the drumbeat. Satan and his sidekicks, the globalists, were behind it all, using COVID restrictions worldwide to attack the church.[22]

Unclean!

Evangelical Christians, as a rule, have no theological problem with modern medicine. Unlike some small sects, most evangelicals will gladly receive lifesaving measures like blood transfusions, antibiotics, and childhood vaccines.

Until the COVID pandemic.

Tim Thompson is the pastor of 412, an evangelical church in Murrieta, California. Thompson is not a medical doctor. Yet, Thompson regularly preached the dangers of the COVID vaccine.

Using Old Testament terminology, Thompson declared the vaccine "unclean."[23] Fearing vaccine mandates in the workplace, Thompson created a downloadable form for a religious exemption.

Thompson was not the only evangelical leader sermonizing medical and legal advice. After posting the caveat "this should not be construed as legal advice," Compass International nonetheless outlined various regulations from the Equal Employment Opportunity Commission, the FDA, and the Federal Trade Commission to be used when seeking a religious exemption for the COVID vaccine.[24]

Why now? Why was a lifesaving, researched, and approved medical treatment declared "unclean" amid a deadly pandemic?

On the word of a few peripheral medical voices, many evangelicals declared the COVID vaccine to be a defiling, unnatural invasion. Once injected, the vaccine alters genes and renders the recipient "transhuman," and as such, recipients are no longer the "sons and daughters of Adam."[25]

Carrie Madej is one such peripheral voice and was a favorite of evangelicals. Madej is a nonpracticing doctor of osteopathic medicine who spends her time "educating others on vaccines, nanotechnology and human rights."[26] Madej equates the mRNA technology used in COVID vaccines with the recombinant DNA technology used in creating genetically modified seeds.[27]

Not the same. At all. But extra scare points for bringing up genetically modified organisms (GMOs), another misunderstood area of science.

RNA is closely related to DNA, the molecule that carries the genetic code. Messenger RNA, or mRNA, is a type of RNA molecule. When our cells need to make a protein, a copy of the DNA gene coding for that protein is made on an mRNA molecule. The mRNA copy leaves the cell nucleus (where the DNA is) and goes out into the cell cytoplasm where protein production occurs.

Think of it this way: grandma's cookbook contains all her recipes. You make a copy of one cookie recipe on a notecard and carry it back home, where you create the cookies. DNA is the cookbook; mRNA is the notecard.

In a process that closely mimics our own biological process, an mRNA molecule with the "recipe" for one of the protein spikes on the COVID virus is made in a lab. The mRNA molecule is then coated in an oily microbubble for protection.

When injected, the oily microbubble fuses with a muscle cell. The cell uses the mRNA "recipe" to make spike proteins. A spike protein alone won't make you sick, but it will teach your body to attack anything presenting the protein—such as an actual COVID virus.

Here are the important take-homes: The mRNA in a COVID vaccine never enters the cell nucleus where your DNA resides. All the protein-making machinery is outside the nucleus, in the cytoplasm. All the raw materials needed to build the spike protein are in the cytoplasm. The spike proteins built in your cells were constructed from raw materials already present in your cells.

And unlike DNA, mRNA is very fragile. The mRNA from the vaccine breaks down and is destroyed within a few days. The mRNA from a vaccine never even gets close to your DNA, much less changes it.

But in an evangelical world primed for science denial, the claim took root.

Secret Knowledge

When my son was a preschooler, we took our dog, Winnie, to the vet to be microchipped. Winnie, being a whippet, ran like the wind and had a bad habit of trying to dart out of the house. My son was far more excited about this mundane bit of pet care than you'd think he would be. As it turns out, my son thought the microchip would allow him to control Winnie's speed and direction, like one of his remote-controlled cars.

Not everyone is so excited about microchip control.

Initially, Anthony Fauci was amused when he heard the rumor that he and Bill Gates had placed a microchip in COVID vaccines as a way to control the masses. As the pandemic wore on and vaccine resistance grew, Fauci was not amused. People believed it.

Fauci and other scientists did not anticipate "such an egregious distortion of reality."[28] Francis Collins found the never-ending stream of conspiracy theories and misinformation affirmed by his fellow evangelicals frustrating, concerning, and dangerous.[29]

Evangelical alarm over vaccine microchips is the latest entry in a history of fears about new scientific technology and the "mark of the beast" from the book of Revelation.[30] In the 1950s and 1960s, Christians feared the technology of the large telephone companies moving into rural areas. The new three-digit area codes were none other than the three-digit number of the beast. Again and again, the mark reared its beastly head: In the 1970s, with credit card numbers. In the 1980s, with bar codes. In the 1990s, with the internet. In the early 2000s, with all sorts of tracking technology. And here we are in the third decade of the twenty-first century with RNA in our vaccines.

In scary and uncertain times, all of us, not just evangelicals, crave answers. We crave certainty. In such an environment, conspiracies take hold.

In the second century, Christianity had no Bible or central structure, or much structure at all beyond a house church, for that matter. The gnostics were a sect that claimed special knowledge, knowledge passed down to them, and only them, by Jesus himself. Only those "in the know" were saved. Like those second-century gnostics, many evangelicals fell for the appeal of secret knowledge. As a member of the enlightened, they feel secure knowing what's "really going on" while everyone else is duped. They are wise and the rest of us are sheeple.

There is a faith over fear that is God-honoring. A faith over fear that is neighbor-loving. A faith that gives thanks for answered prayers in a pandemic in the form of a science-based vaccine.

There is a faith that makes room for trusting science.

6

Life in the Bubble

I have a clipping from a Depression-era newspaper, sepia-toned with age. It's an advertisement for an upcoming "Gospel Meeting," more commonly known as a *revival* outside the church tradition in which I grew up. "Come, bring your friends!" "No collections taken!" Every evening for ten days. *Ten*. Restorationists are not here to play.

The ad features a photo of the guest evangelist, young and earnest and dressed in his best suit and tie: my Granddad Kellogg.

Interestingly, the hosting congregation was the first iteration of the very church where I am now a member. What would my granddad think of the twenty-first-century version? I don't know what would shock him more—the drum set on stage or the women!

As newlyweds, my grandparents were gifted seven acres to farm in central Texas. They were faithful churchgoers, and as Providence would have it, my grandfather was asked to preach one Sunday. It was a very short sermon, recalled my grandmother, but apparently it was a hit.

Encouraged by the brethren to follow the calling, my grandparents sold their few possessions, packed up the baby, and moved to Abilene, Texas, for the summer.

My grandfather had just enough money for a few Bible classes at Abilene Christian College. Thus was launched his preaching career.

My grandfather was an early adopter of the multimedia presentation. With a black marker and one of my grandmother's good white bed sheets, he would outline a sermon and tack it up at the front of the auditorium. And there you go—DIY PowerPoint. Spending the night at my grandparents' parsonage meant grandkids on the floor on a pallet of bedclothes, falling asleep on the Steps of Salvation.

My grandfather's minimal preparation for the pulpit was not unusual for the time, and not terribly far out of the norm for evangelicals of the twentieth and twenty-first centuries. Too much education could be seen as a negative—zeal for the gospel trumps any highfalutin credentials.

And yes, highfalutin is to be avoided.

John Piper is a dominant name in twenty-first-century Reformed theology. Piper is the founder and leader of Desiring God, an international web ministry with millions of monthly users. It is hard to find an evangelical church untouched by Piper's influence or the influence of other neo-Reformed leaders and teachers.

Here's Piper: "My guess is that some of the most fruitful pastors in America have nothing but a high school education. Can I name one? C. J. Mahaney. . . . And C. J. Mahaney founded a Pastor's college. Not a big highfalutin one. He just knows that you need some . . . need some what? Not schooling. You need some insight, some wisdom, some shaping, some depth, some Bible, some experience. All of those are biblical. So yes, Amen, let's not elevate levels of schooling."[1]

A Perfect Storm

There is an unfortunate thread of anti-intellectualism running throughout the history of the evangelical movement. We have an instinctual aversion to experts, especially if the experts are championing ideas not considered common sense to most people. We've been raised on "plain readings" of Scripture and the wisdom of everyman. And everyman, of course, resides in twenty-first century Western evangelicalism.

Despite the many examples of highly intellectual, highly educated evangelical leaders and lay people, a culture of valuing personal feelings over the intellect percolates through evangelicalism. It is seen in oft-repeated platitudes like "Just give me that old-time religion" and "All you need is a simple faith." Contemporary Christian praise songs frequently elevate feelings over intellect with lyrics like "I don't need to see it to believe it"[2] and "When I trust You I don't need to understand."[3]

It's not that feelings and emotions are valueless; it's that the intellect is *de*valued.

It's not for nothing that evangelicals own the quippy catchphrase "Christianity is a relationship, not a religion." As evangelicals diverged from generic Protestantism, they embraced the personal: A personal Lord and Savior. A personal conversion experience. A personal walk with Jesus. And supremely important, a personal revelation—a Bible inerrant in all it claims about the nature of God and the nature of the world.[4]

Initially, the dividing line between evangelicals and mainline Protestants was rather blurry. But all that changed following the Scopes monkey trial.

The Scopes trial was a perfect storm—two monumental intellectual movements, both birthed in the nineteenth century, colliding in the twentieth century.

The first crest of the storm was historical biblical criticism. Emerging from nineteenth-century German theologians, historical biblical criticism spilled out of the academy and into the churches in the twentieth century.

Historical biblical criticism is not criticism as in "disparagement" or "disapproval." *Criticism* in this case means to explain, to evaluate, to discern. As objectively as possible, scholars attempt to identify the meaning intended by biblical writers. Historical criticism is a way to peer into the backstory of the Bible the way a journalist pokes around for the backstory of a breaking news event.[5]

The goal of critical investigation was not to falsify the Bible but to understand it better. Historical criticism tells us that the Bible is not a book dropped into our laps, made of whole cloth. Historical criticism tells us that the Bible is marked with human fingerprints: cultural fingerprints, genre fingerprints, linguistic fingerprints, historical and archaeological fingerprints.

But the very suggestion that things were not as they appear to be sent waves of angst across the evangelical world. Suggesting, for example, that the first eleven chapters of Genesis are not a literal and historical account of the beginning was unsettling.

Intellectuals in their ivory towers were the source of the trouble. Intellectuals took an inerrant Bible that says what it says and means what it means and fractured it.

And the angst persists, unabated, into the twenty-first century.

In October 1978, more than two hundred evangelical leaders met in Chicago and drafted the Chicago Statement on Biblical Inerrancy.[6] The Chicago Statement declares the Bible to be "without error or fault in all its teaching, no less in what it states about God's *acts in creation* [italics mine], about the events of world history . . . than in its witness to

God's saving grace." In the Chicago Statement, rejection of inerrancy is declared to be a refusal of the "true Christian faith." The Chicago Statement remains a litmus test and a contractual requirement in many evangelical seminaries, colleges, and churches.

When the book *The Making of Biblical Womanhood* was published, Denny Burk (of the Council on Biblical Manhood and Womanhood) publicly threw down the Twitter gauntlet to author and Baylor University professor Beth Allison Barr: "Do you affirm the doctrine of inerrancy as articulated in the Chicago Statement?"[7]

The first crest in the perfect storm, historical biblical criticism, dismantled the inerrant Bible of the evangelical movement.

Adding fuel to fire, here comes Charles Darwin, riding the wave of the second crest. As academics were undermining the Genesis account of a young earth and an instantaneous special creation, Charles Darwin showed up with a new way to explain human origins. And humans, no longer specially created, were no longer the center of it all, the apple of God's eye.

In the heyday of the scientific method, evangelicals spurned science, chose emotion over reason, and set the movement on a path of anti-intellectualism.

So much for a personal savior.

In the perfect storm of inerrancy denial and evolution acceptance, evangelicals saw intellectual elites attacking the very core of evangelical faith. The battle crescendoed in 1925 in a courtroom in Dayton, Tennessee: the Scopes monkey trial. In the "trial of the century," Scopes was convicted, and anti-evolution won the day. In the heyday of the scientific method, evangelicals spurned science, chose emotion over reason, and set the movement on a path of anti-intellectualism.

Intellectual Isolation

The Scopes trial was a fork in the road for Protestants. From that point on, evangelicals forged their own separate journey. William Jennings Bryan, champion orator, winning Scopes attorney, and the hero of anti-evolutionists, set the course: evolution and modern biblical criticism weren't simply intellectual concepts but ideas that threatened both civilization and religion.[8]

Out on their own, evangelicals built an elaborate subculture. They built religious schools, seminaries, Bible colleges, and publishing houses. They created women's and men's ministries, youth ministries, and Bible study organizations. A vast media empire arose: radio stations, television stations, production companies, and recording labels.

Evangelical parachurch groups proliferated and still flourish—groups like Young Life, Campus Crusade for Christ (now Cru), and Fellowship of Christian Athletes. Other Christian groups also field their own parachurch organizations, but evangelicals by far dominate the category. Evangelicals organized church-based mother's-day-out programs, elementary schools, and secondary schools. It is possible to progress from preschool to graduate school and never leave an evangelical school.

We listen to our own, we read our own, we study our own. The evangelical subculture is so large and multifaceted we have no problem doing so.

The size and success of the evangelical subculture breeds intellectual isolation.[9] Evangelical authors usually write for one of the many successful evangelical publishing houses, journals, or media outlets. There is no need to engage an audience who denies God or faith or even a non-evangelical worldview, for that matter. Evangelical content creators do just fine preaching to the choir, thank you very much.

In an (admittedly) extreme example of tribalism, my growing-up church hesitated (big time) before showing James Dobson films. We were given a verbal asterisk from the pulpit—Dobson does not agree with us regarding baptism, so we should pay attention only to his teaching on home and family. That was the seventies, but even today, churches are jettisoning their beloved Beth Moore Bible studies because Moore stepped outside the evangelical circle.[10]

Living inside the (really large) evangelical bubble, we never have our presumptions challenged. Working inside the bubble, evangelical content producers never have to engage non-evangelical positions from the *actual* position-holders *themselves*.

Oh, we have apologetics. We can refute and rebut with the best of them. And in doing so, we argue against our *interpretations* of what these positions are. We build a straw man, tear him down, and think we've done the work of looking outside our camp.

Teaching and Training

"Robert, are you crazy? Honey, you don't know anything about a school. You're not an educator. You can't found a school."

Jones answered, "I know I can't. But God can."[11]

And just like that, Bob Jones College (now University) began educating young adults in a "spiritually safe" environment, free and far away from the corruption of Darwin and his ilk. Importantly, Bob Jones College would *not* be the education of a Harvard or Yale or Princeton, schools that had abandoned their Christian heritage and sold their souls for academic excellence.

Jones was not the only evangelical leader distrustful of higher education. Oral Roberts, Jerry Falwell, and Pat Robertson, none of whom had educational credentials, likewise started their own evangelical Christian universities. Others followed suit. Yale, Harvard, and Princeton were our cautionary tales. Evangelicals needed universities free from egg-headed scholars poking their noses into long-settled facts of biblical inerrancy and special creation. Regardless of whether a student was studying pre-med or accounting or was preparing for ministry, evangelicals wanted institutions of higher education free from the poison of intellectuals in liberal ivory towers.

It wasn't long before evangelical universities were also suspect—the two primary destinations of the slippery slope being theology and science. Answers in Genesis gets right to the point: "If they (colleges) are willing to take a stand on a literal Genesis, that's a great sign they hold firmly to biblical authority. . . . If the school refuses to take a position or, worse, takes the position of evolution and/or millions of years, you can see that they allow other things, such as secular scientists' interpretations, to be the authority over God's Word—and that's dangerous."[12] Bible colleges, schools of preaching, and seminaries grew up in the shadow of distrust in evangelical universities. Evangelicals needed places to train ministers, teachers, and leaders where inerrancy and special creation weren't questioned.

Bible colleges are not research institutions. There is no need to dig into research in biblical criticism or to acknowledge past or current research in evolution biology. Anti-evolution discussions and apologetics resources are commonplace in Bible college science classes. For example, in Moody Bible Institute's generic "life science" course, three required resources are listed in the syllabus: the textbook, a Bible, and the full-length creationist film *Is Genesis History?*[13]

Although the primary objective of a Bible college is to train students for various types of ministry, some offer a limited number of other ministry-related degrees, usually teaching, music, or general business.

"Schools of preaching" are unique to the Churches of Christ and have trained thousands over the last few decades. Like a Bible college, they exist to train ministers. Schools of preaching offer certificates of completion after two years of study or less. Women are not accepted to preaching tracks but may take classes for ministers' wives or nonpreaching ministries. Like at a Bible college, inerrancy and special creation are the default positions.

Schools of preaching emerged in response to the dangers of "liberalism" in Church of Christ–affiliated universities. Schools of preaching are historically adamant in their anti-intellectualism. Harvard-educated professors in affiliated universities were called "Harvard specialists" instead of "preachers" or "Bible teachers" by supporters of schools of preaching.[14]

The autonomous theological seminary, separate from a college or university, is an American creation.[15] In seminaries, students study for advanced degrees in Bible, theology, or other areas of ministry. Seminaries are often under the direct control of a specific denomination. Seminaries, large and small, have trained evangelical leaders, teachers, and preachers for more than a century.

Two of the largest and most influential evangelical seminaries are Dallas Theological Seminary and Southern Baptist Theological Seminary in Louisville, Kentucky. Both seminaries are bastions of conservative evangelical thought. Biblical inerrancy, young-earth creationism, and traditional gender roles are unquestioned.[16]

Like Bible colleges and schools of preaching, seminaries educate ministry students in isolation from the rest of academia. Teachers and professors in these institutions research, study, write, and teach in isolation from scholarly thinking in all other academic fields. Historically, theological training was done in the context of a university: theologians, scientists, and other academics studied in the same facilities.[17] The university-based model is still the norm in Europe.

Training evangelical leaders, teachers, and preachers in academic exile preserves doctrinal purity, a priority for most Bible colleges, seminaries, and schools of preaching. But an insular education comes at a cost. The price is a loss of "cross fertilization" between theological thinking and thinking in other academic fields—including the sciences.[18]

An education in isolation provides no opportunity for scholarship in science to inform theology. It is no surprise, then, that evolution denial and a young earth are the norm in evangelical Bible colleges, preaching schools, and seminaries. The price we pay is perpetuating anti-intellectual, antiscience mindsets in our leaders, teachers, pastors, and preachers.

In a growing number of Christian universities and colleges, however, professors in the theology school teach and research alongside professors doing real research in science. And in a growing number of Christian colleges and universities, evolution is taught as the foundation of biology. Notable are the biology departments of Baylor University (Southern Baptist), Abilene Christian University (Church of Christ), and Calvin University (Christian Reformed Church).

In its departmental statement on evolution, Baylor University is upfront: "We are a science department, so we do not teach alternative hypotheses or philosophically deduced theories that cannot be tested rigorously."[19] At Abilene Christian University, evolution is integrated into the coursework of biology majors from day one. And in the biology departments at each of these three Christian universities, faith informs education in biology:

- "We have a Biblical mandate to understand God's world."—Baylor University[20]
- "Everything we do arises from this Christian perspective."—Abilene Christian University[21]
- "We believe God brings forth creation through an evolutionary means."—Calvin University[22]

Not long ago, I spoke at a science and faith conference at Abilene Christian University, my alma mater. I was invited to an informal lunch with a small group of biology professors and Bible professors.

Theologians and scientists, eating sandwiches and discussing contemporary issues of the day, as fellow researchers and colleagues.

Amazing.

Reverse Snobbism

New York Times best-selling author and evangelical pastor Tim Keller sees "reverse snobbism" in evangelical anti-intellectualism.[23]

Sometimes, all it takes for a source to be dismissed is the implication of higher education.

Russ Miller is the author of creationist books, and he owns a creationist Grand Canyon river rafting company. Miller's keynote presentation at the 2019 Steeling the Mind conference quotes eminent biologist Eugenie Scott on abiogenesis, a topic in biology about which we are still learning.[24] Scott describes what we already know about abiogenesis but adds, "There's a lot of iffy stuff in the middle."

Miller quotes Scott in his keynote, but adds a snarky caveat: ". . . with a lot of iffy stuff in the middle? That's the modern 'college' explanation of how life got started without God."[25] No rebuttal. No refuting facts. Miller debunks abiogenesis with a single word: "college."

Miller is not in academics, but John MacArthur is. *Christianity Today* calls John MacArthur one of the most influential preachers of our time. The *MacArthur Study Bible* has sold more than one million copies. In a recent interview, MacArthur was asked to comment on Francis Collins, a world-renowned geneticist, former head of the NIH, and an upfront, outspoken, and enthusiastic evangelical Christian who accepts the evidence for evolution and the age of the earth.

Here's MacArthur: "They [scientists like Collins] are so elevated in their own minds as to the importance of the role they play. . . . They think that if Christians don't bow to this elite, scholastic, scientific community, we're going to look like fools."[26] This is from a man who leads a Christian university and a seminary.

Setting a Course

In the race to space decade of the 1960s, the United States was pouring a fortune into science education. To be anti-science was un-American. Instead of simply fighting evolution in schools, evangelicals crafted a new field of study meant to stand on equal footing with evolution in the classroom: *creation science*. Creation science was presented as an alternative scientific model, equivalent in standing to the evolution model.[27]

And when the US Supreme Court ruled in 1987 that creation science is a religious doctrine and cannot be taught in public schools, we retooled creationism. We spiffed up creation science with technical terms, acquiesced to an old earth, and introduced a brand-new model to rival evolution: *intelligent design*.

Creation science and intelligent design are not, by their nature and definition, subject to investigation by the scientific method. Instead of researching, writing, and creating new knowledge, evangelicals redefined science. In wedding the terms *creation* and *science*, evangelicals intuitively understood the authority of science, even as they rejected it.[28]

In the early twentieth century, evolution was the line in the sand. Evangelicals stood on one side of the line with a death-grip on a literal reading of Genesis in one hand and a seven-day, special creation of a six- to ten-thousand-year-old earth in the other hand.

On the other side of the line stood modern science.

Darwin's big idea still holds in the twenty-first century. It is the bedrock of modern biology, genetics, medicine, agriculture, and conservation—all of life science, actually. Alongside evolution stand modern physics and chemistry and geology, testifying to an ancient earth.

When evangelicals declared war on evolution, they set a course for science denial in the modern age.

The Scopes trial changed the way evangelicals talked about science. It changed the way we talked about scientists. And it changed the way we thought, talked, and responded to research evidence and to the academics doing the research.

The irony cannot be missed. In a time when evangelicals were founding universities and colleges, evangelicals—even highly educated evangelicals—were rejecting intellectualism and were devaluing the life of the mind, particularly in the sciences. The legacy of anti-intellectualism in science has left its mark.

Evangelical scientists working in industry or in research often live dual lives—science is a vocation, faith is religion, and never the two shall meet. But for many evangelical scientists working in Christian colleges and in some Christian universities, the situation is precarious. The way to get along and keep your job is to stay quiet about controversial issues. The result? Few are intentionally studying science in relationship to theology in an evangelical context.[29]

There's another scenario that plays out among scientists in the evangelical world. If Genesis is read as a literal scientific and historical account of beginnings, we immediately have a problem with modern genetics, biology, chemistry, physics, and geology.

But doesn't the apostle Paul tell us that God chooses the foolish things of the world to shame the wise? When science and faith are in conflict, evangelicals sometimes choose to play the fool for Christ.[30] Georgia Purdom is a molecular geneticist. Nathaniel Jeanson is a cell biologist. Kurt Wise is a paleontologist. All three have PhDs from stellar universities—Purdom from Ohio State, Jeanson and Wise from Harvard. And all three are young-earth creationists. All three believe science supports a six-to-ten-thousand-year-old universe and instantaneous creation. All three acknowledge that their interpretation of scientific evidence is far outside the norm.

Purdom, Jeanson, and Wise proudly stand against the vast majority of the world's scientists. Any conflict between scientific evidence and the Bible is awarded to the Bible. When evidence is viewed through a biblical lens, they say, the conclusion is always a young earth and instantaneous creation.

The intelligent design version of creationism likewise has a few leaders with impressive degrees. While intelligent design advocates accept an ancient universe, they reject evolution, the foundation of modern biology.

For decades, evangelicals have been told that the universe is less than ten thousand years old. For decades, evangelicals have been told that life appeared instantaneously, more or less in the forms we see today. For almost a century, we have undermined how evangelicals look at biology and geology.

In doing so, we have undermined the ability of evangelicals to look at the natural world and understand it.

And in the twenty-first century, when faced with discerning other elements of the natural world like climate and disease, evangelicals are suspicious of the evidence and suspicious of the experts.

We've been conditioned: when in doubt or when something challenges long-held beliefs, the true Christian rejects the intellect and elevates faith.

7

You and Me against the World

It is a flawless day at the State Fair of Texas, the kind you only get in a Texas October. Surrounded by sunshine and family, the guy smiles for a selfie in front of Big Tex and posts it on social media.

The scene is postcard pretty . . . until your eyes pan down to his T-shirt. It is a parody of the famous "March of Progress" illustration. You know the one—a single file, one-in-front-of-the-other march starting with a monkey, progressing through various ape-men, and finally culminating in an upright modern human.

In the T-shirt version of the march, the modern human is positioned at the starting point. At each step toward the finish line, the human crouches further and further to the ground, until at last, he is a woolly, four-legged, barnyard animal . . . a sheep.

The image is curious, but the caption is jarring: "I HATE SHEEPLE." (To Google for various renditions of this image.)

We've got a few things going on here with this T-shirt.

For a couple of years during the COVID pandemic, some of my evangelical social media follows took a break from posting Christian thoughts-for-the-day and instead posted their hot takes on all things COVID.

In the "March of Progress" parody, we have an unflattering reference to evolution. In the reference to "sheeple," we have an unflattering reference to those who hold a majority opinion. In the choice of the verb "hate," we have disdain, no, actual *hate*, for those with whom we disagree. And in the context of this social media post, it is those who side with conventional science in the pandemic.

And finally, the switch from inspirational Christian posts to memes opposing a popular position reveals a decades-old evangelical perspective on science.

Cheering for the Underdog

Evangelicals are steeped in a "David versus Goliath" mindset, and we are *always* David. We are always Moses, never the Egyptians. We are always Elijah, never the prophets of Baal. We assume the underdog position is the noblest. We assume the underdog is always right and the crowd is always wrong. We lionize standing against the majority. We don't tell stories about the times the majority was right.

Evangelicals love an underdog story, especially a faith-based one. Extra points if it is about sports. The faith-based feature film *American Underdog* is based on NFL star Kurt Warner's memoir *All Things Possible: My Story of Faith, Football, and the First Miracle Season*.

Award-winning religion and culture writer Jonathan Merritt sees American evangelicalism as "predicated on the existence of an enemy to fight."[1] Over the years, evangelicals have fought many enemies, but no enemy has drawn as much evangelical fire or for as long as science, and specifically, the science of evolution.

When it comes to science, evangelicals have been told for years that it's a fight and we are the underdogs. And not simply a fight, it is a battle. An all-out war.

Nowhere is the science battlefield metaphor more apparent than on college campuses. Answers in Genesis has a "how to send your kid off to college" book with the ominous title *Already Compromised.* The problem with American colleges, even Christian colleges, is science. Chapter 2 is titled "Welcome to the War," and it gets straight to the point: "It is spiritual warfare. . . . At the core of all these so-called scientists and educators is the commitment to Darwinism (with its tenets evolution and millions of years)."[2]

Popular evangelical films also love the war-on-campus theme. The first movie in the *God's Not Dead* trilogy features a smug atheist professor who publicly humiliates a student for denying evolution. The professor's nefarious goal is to destroy the faith of any student brave enough to stand up for God.

It's indeed a war, and the battleground is our very culture.

Science and the Culture Wars

He's wearing jeans, a plaid shirt, a trucker hat, and sturdy work boots—definitely a lumberjack vibe. He's standing on "solid ground" and he's wielding an ax of "truth." This cartoon illustration comes to us from Creation, Evolution, and Science Ministries and was featured in a keynote at the 2019 Steeling the Mind Bible Conference.

A tree is growing out of soil labeled "old-earth beliefs." Like the scary trees in *The Wizard of Oz*, this tree has a sinister "face" and gnarly branches twisting out and away from its base. The trunk giving rise to this arboreal monstrosity is labeled with one word: "Darwinism."

Growing from the trunk of "Darwinism" (a.k.a. evolution) are branches of abortion, false religions, racism, humanism, socialism, communism, and satanism.

The image is not subtle. It's like (cue the marching band) "Evolution starts with *E*, and that rhymes with *C*, and that stands for communism!" And we got trouble, right here in River City.

In the 1970s, the American anti-evolution movement was gaining steam. Henry Morris, along with Tim LaHaye (later the author of the wildly popular *Left Behind* books), established the Creation Science Research Center. Morris and LaHaye set out to strip creationism of explicitly religious language and embed creation science in public schools as an alternative to evolution science.

When research in "creation science" stalled, anti-evolution activists turned their sights on culture. Advocates found it easier to argue that evolution was the source of cultural decay than to argue that evolution science was false.[3]

Before long, the Creation Science Research Center was fighting not only evolution but also sex education, abortion, women's rights, and gay rights. The Center claimed a link between evolution acceptance and divorce, abortion, and rampant venereal disease.[4]

Advocates found it easier to argue that evolution was the source of cultural decay than to argue that evolution science was false.

Creationist organizations and evangelical churches weren't alone in blaming evolution for the breakdown of culture. On the floor of the United States House of Representatives, former Majority Leader Tom DeLay blamed the Columbine school shooting not on guns but on school systems that teach children that "they are nothing but glorified apes who have evolutionized out of some primordial soup of mud."[5]

Evolution opponents are not opposed to pulling out all the emotional stops when connecting science to the downfall of culture. In its review of the PBS series *Evolution*, the

Dallas-based ICR equates those who teach evolution to the 9/11 hijackers: "Militants in our schools desire to alter the life and thinking of our nation and are no better than those who attacked our country."[6]

Driving a Wedge through Culture

The Discovery Institute is the primary think tank for the more sciencey-sounding manifestation of creationism, intelligent design. Intelligent design accepts an ancient age for the universe but rejects evolution. The Discovery Institute is large and well funded and publishes lots of books. As of this writing, four of the top ten Amazon bestsellers in the "creationism" category are by Stephen Meyer, a Discovery Institute Senior Fellow.

In 1999, a jaw-dropping document was leaked from the Discovery Institute. The "Wedge Document" was a manifesto laying out long-term plans to remove evolution science from public consciousness and replace it with a Christian "version" of science.

Seriously.

Initially, the Discovery Institute was silent regarding the leak. They eventually owned up to it and published *The "Wedge Document": "So What?"*[7]

The "Wedge Document" never identifies evolution science by name. Throughout the document, the terms *materialism* and *materialistic* are used as euphemisms for evolution. Without a doubt, however, "materialism" refers to evolution. The resources listed at the conclusion of the document erase any lingering uncertainty—the enemy to be fought is evolution.

Five-year plans and twenty-year plans outline goals intended to reverse the dominance of the "materialist worldview" and to replace it with a science that reflects Christian convictions.

The targets for such a reversal are not just public schools but society at large. Strategies to achieve this included changing public policies, publishing op-ed pieces, and engaging cultural influencers for the cause.

But why? Why do we need to overthrow evolution and install a Christian "version" of science?

The introductory section of the document is unequivocal. It's a culture war: "This materialistic [read: evolution-affirming] conception of reality eventually infected virtually every area of our culture, from politics and economics to literature and art. . . . Discovery Institute's Center for the Renewal of Science and Culture seeks nothing less than the overthrow of materialism and its cultural legacies."[8]

The Culture War Goes on Vacation

If the messages of cultural doom are just too much and the covert nature of the "Wedge Document" feels a bit too creepy, you can lighten up a bit at a popular tourist venue.

The Creation Museum in Petersburg, Kentucky, along with its sister venue the Ark Encounter, draws hundreds of thousands of visitors each year. The museum features botanical gardens, a petting zoo, a special effects theater, and all sorts of exhibits, speakers, overnights, and of course, dinosaurs.

It's all sunshine and puppies until you turn the corner.

Welcome to Damnation Alley. It's dark and dilapidated, greasy, rundown, graffitied, and scary. We're obviously on the bad side of town. Seedy signs and seedy lighting set the mood.

Watch out! A giant wrecking ball engraved with "MILLIONS OF YEARS" smashes into a church, scattering the bricks. A scientist stands in the background, gathering the bricks into a wheelbarrow.

In nearby houses, we see kids playing violent video games, watching porn, smoking weed, and lamenting a teen pregnancy.

Exit to the gift shop where you can, if you're lucky, purchase a DVD titled *The Cure for a Culture in Crisis: It Doesn't Take a Ph.D.* From this DVD you'll learn that evolution is the cause of the breakdown of society and that "people with PhDs come to the most dumbest [sic] conclusions."[9]

As I write, John MacArthur's Truth Matters Conference just wrapped up its 2022 event. The 2,500-seat auditorium at the Ark Encounter was host for this year's conference. The venue was packed for three days of lectures, panel discussions, and worship. Headlined by MacArthur (who famously told Beth Moore to "go home" at the 2019 conference), the full lineup included evangelical heavy hitters like Owen Strachan, Justin Peters, Ken Ham, and more.

"Recovering a Biblical Worldview" was the theme. The conference website gives just a taste of the content: gender identity, critical race theory, social justice, and above all, the destruction of culture by fads of worldly wisdom.[10] At the close of the conference, Strachan took to Twitter to sum up the event. Two hundred eighty characters could hardly contain his enthusiasm: "Stop doing theology by a hermeneutic of shame. Stop being embarrassed by the Bible."[11]

And what does that look like? Strachan lists several examples of potential embarrassment: a literal Adam, a talking snake, and six twenty-four-hour days of creation. "I am not ashamed of six-day biblical creationism," posts Strachan, "and nothing could make me ashamed of it." Three days of warnings about cultural decay, all summed up with one idea: evolution science is the root of it all.

In tourist venues and at special conferences, in think tanks, in church pulpits, and in the bully pulpits of local school boards, state halls, and Congress, the theme repeats: Christians are facing a full-scale cultural war, and science is leading the attack.

Hidden Agenda

About 113 million years ago, a couple of dinosaurs made their way across North Central Texas, just south of present-day Fort Worth. One was a theropod, a cousin to *Tyrannosaurus rex*, who walked upright on his hind feet. The other was an enormous sauropod, a long-necked plant-eater who left pothole-sized footprints. We don't know if they were buddies or not, but both left their marks in the mud in the Paluxy River basin.

In 1909, teenager George Adams found the fossilized tracks of these two dinos near Glen Rose, Texas. George told his school principal about his discovery, and before long, the tracks were authenticated by the Smithsonian and the American Museum of Natural History.

Glen Rose has a population of less than three thousand, but ninety thousand people visit Dinosaur Valley State Park each year to see the famous footprints. Not everyone comes to see the dinosaur prints, however, at least not the dinosaur prints *alone*.

Many locals claim that human footprints are also present, walking alongside the dinosaurs. In the 1980s, a minister moved to town and opened the Creation Evidence Museum. I've been there and I have the postcard to prove it. Come for the dinosaurs. Stay for the Creation Museum.

Creationists flock to the town, hoping to "prove" a young earth and the coexistence of humans and dinosaurs, having both been created on the sixth day.

In her late eighties, Mary Adams, niece of George Adams (the Glen Rose teen who originally found the dinosaur tracks) spoke to the youth group at the local First Baptist Church. Mary issued a dire warning: don't believe in evolution. If you do, you'll become an atheist. Here's Mary: "If we were not created by God, there's no one to whom we are accountable. We can live exactly as we please."[12]

From the time we could toddle to our own Vacation Bible School class, we were at war . . . at least we sang about it. We marched in the infantry and rode in the cavalry and shot the artillery and flew o'er the enemy because we were in the "Lord's Arrr—meeee!" We jumped into an iron-age vehicle of war and rolled the gospel chariot along. We rolled right over the devil, and then we told him to sit on a tack.

A constant sense of faith under assault is inescapable in the evangelical narrative. And no aspect of faith is more in the crosshairs than our very belief in God.

When it comes to the science of evolution, evangelicals have been told that there is an agenda. The agenda of science is not *really* research, testing, and evidence. That's just what they *want* you to think.

The unspoken agenda of science is atheism.

The Clergy Letter Project, founded by Michael Zimmerman, is a collection of resources for clergy (or anyone) who want to learn more about the compatibility of science and faith. As of May 2022, more than seventeen thousand clergy have signed on to the project, the majority of whom are Christian.[13]

Each year, the project sponsors an "Evolution Weekend" featuring discussions and lectures across the country. The goal of the weekend is to demonstrate that science does not threaten, demean, or diminish belief in God.

Not so! says Ken Ham. According to Ham, the objective of the weekend is nothing of the sort: "Really, it's an attempt to get Christians to compromise God's Word with aspects of the religion of naturalism and atheism. . . . We must recognize evolution for what it is—not a 'proven' scientific idea, as some would mistakenly claim, but an anti-God, anti-Bible ideology imposed on the evidence."[14]

John MacArthur agrees with Ham—evolution is

atheism. To MacArthur, evolution is nothing more than a made-up, no strings attached, get-out-jail-free card: "To put it simply, evolution was invented in order to eliminate the God of Genesis. . . . Society has embraced evolution with such enthusiasm because people imagine that it eliminates the Judge and leaves them free to do whatever they want without guilt and without consequences."[15]

MacArthur is old guard, but Eric Hovind is the current fresh face of creationism. On an episode of his show, *Creation Today*, Hovind and his guests discuss the question "Why do people believe evolution?"[16]

The answer? People don't want to accept God. And if you don't want God, you must find something else to believe. Simple as that.

Creation Ministries International likewise portrays evolution as nothing more than a rebellious, God's-not-the-boss-of-me approach to life: "The concept of an overarching Creator to whom we must be accountable is unthinkable and unacceptable. So, although one can be bombarded with the supposed supporting evidence for evolution, it pays to understand that's really not where the battle is."[17]

For many everyday evangelicals like Mary Adams of Glen Rose, as well as influential evangelical leaders like Ham and MacArthur and Hovind, there is no evidence for evolution. None. Evolution is simply something scientists invented because they want us to reject God.

The Story We Tell

In the evangelical world, evolution is, by default, atheistic.

Technically, evolution is agnostic. Describing it as "agnostic" does not convey a negative connotation but simply that the mechanics of biological evolution can be explained without referencing God—the same way we can explain the mechanics of the water cycle or the mechanics of childbirth without referencing God.

Although the theory of evolution says absolutely nothing about God or religion or faith or any other worldview for that matter, it's not the story we tell.

An extensive study published in 2020 by professors and researchers in higher education is particularly eye-opening. The researchers are not theologians, nor are they professionally associated with any religion.[18] Over one thousand college biology students (both majors and nonmajors) across multiple large public research-intensive universities participated in the study. The researchers found that among self-identified religious students, 49 percent believe that evolution *requires* atheism, which probably doesn't surprise you. But this surprised me: 47 percent of nonreligious students *also* believe that evolution requires atheism.

Evangelicals have done such a good job coupling evolution with atheism that even *non*religious people think the same.

Save Us, Uncle Sam

Americans, and especially evangelical Americans, have been fostered since childhood in rights and freedoms. And no right is more fundamental than religious freedom.

If your culture is under attack by science, even your very belief in God, where do you turn for help?

We look for sympathetic ears in positions of power. We adulterate science with politics. And when we do, we often make opposition to science a political position.

When Dan Patrick, the powerful lieutenant governor of Texas, addressed the 2021 CPAC, he described the fight against liberal causes as a matter of "darkness and light," a matter of "goodness and evil."[19] It's no wonder evangelicals often see politicians as allies in a culture war when politicians blatantly evoke religious imagery to make a political point.

Adulteration of science with politics is not new. Longtime activist Phyllis Schlafly has fought against political positions she sees as opposed to Christian culture since the 1970s. To Schlafly, evolution is not a theory foundational to modern biology, it is a political weapon of the left: "Liberals see the political value to teaching evolution in school, as it makes teachers and children think they are no more special than animals."[20]

In the 2004 presidential election, House Majority Leader Tom DeLay made opposition to evolution a major theme of the campaign. To DeLay, evolution was a political accusation to be used against liberal judges, teachers' unions, and anyone who supported teaching evolution in schools.[21]

A year later, Missouri state representative Cynthia Davis once again invoked the 9/11 hijackers. Just like the hijackers, Davis said, liberals are taking our country somewhere we don't want to go when they teach evolution.[22]

Politicians aren't the only ones making science political. Ann Coulter is a Fox News regular and an outspoken political pundit. Her best-selling book *Godless: The Church of Liberalism* is, without a doubt, a political book. Yet, Coulter devotes eighty-two pages to evolution. Coulter calls evolution "liberals' creation myth" and "about one notch above Scientology in scientific rigor."[23]

When we speak about evidence-based, peer-reviewed, well-established science in the language of politics, we have a problem. When evangelicals use politics to fight a culture war with science, we lose the ability to discern between politics and science.

Science Is the Enemy

In arguably the freest nation on the planet, evangelicals see themselves as a disrespected minority, afflicted, perplexed, and persecuted. We are fighting a culture war. And although the culture war has many fronts, evangelicals have been told for decades that evolution science is the root cause of a myriad of evils.

As the twenty-first century progresses, it is clear: evolution is not going away. Modern genetics sealed the deal. In agriculture, medicine, conservation, epidemiology—all biology, actually—evolution theory is foundational.

Evolution theory, like all scientific theories, is clarified and tweaked as science rolls on, but it is not going to be scrapped. Despite dire predictions by creationists, evolution theory is not going to end up in the Disproven Dumpster alongside a flat earth, a geocentric universe, and disease-causing foul air.

In a culture war, it is much easier to blame evolution for violence and atheism and even disrespectful children than it is to prove evolution false.

In the war against evolution, evangelicals stand against a science that is as foundational and widely accepted as germ theory, gravitation theory, and atomic theory. Unquestionably, evangelicals are standing up to the crowd.

We lionize the underdog. We fill with pride when we stand against the majority. But what if the majority position is the correct one? What if the crowd is right?

When faced with twenty-first-century scientific problems like climate change and pandemics, we take up battle stations. Evangelicals have been conditioned: we are suspicious of the evidence, suspicious of the experts, suspicious of the majority position.

There's a popular evangelical meme that crops up occasionally. It features a mom (sometimes a dad) kneeling on the ground, holding an open Bible in one hand and holding a shield in the other. In front of mom, seated on the ground, is a child. The child, oblivious to all around, is smiling at mom and the Bible. Mom weakly smiles back.

Despite the smiles, all is not well. Mom's back is bloodied from multiple flaming arrows shot into her back. Her body protects the front of the child, while the shield in her hand protects the child from arrows flying in from behind him. The moral of the meme is obvious: parents protect their children from the flaming arrows of the enemy with their bodies and the Bible.

And . . . there's our battle theme.

But wait. The 2021 version of the meme has been updated. In addition to flaming arrows protruding from mom's back and from the shield, there are also multiple syringes. In 2021, evangelicals fought a new enemy: COVID vaccines.

In the fight against evolution, we've been carefully trained. We've been told it's a war and we are up against Goliath. We've been told our morals and our beliefs are under attack. We've gravitated to political ideology that feeds this narrative. We are battle-hardened.

And we're ready to fight the next enemy science throws at us.

8

They'll Know We Are Christians by Our Constitutional Rights

Mary Mallon was proud of her craft. At the turn of the twentieth century, Mary was in demand by wealthy Manhattan families for her skills as a cook. One day, a city official knocked on the door of the fancy brownstone where Mary was employed. He had a few simple requests to make of Mary. Could he please have a sample of her blood, her urine, and her feces? Oh, and, this little question: Do you regularly wash your hands after going to the bathroom?

That was one question too many for Mary. She grabbed a carving fork, swore, and lunged for the man, who wisely exited the home.

Four months prior to his abrupt ouster from Mary's place of employment, George Soper, a New York City sanitary engineer, was asked to investigate an epidemic that would eventually sicken thousands of New Yorkers. Soper began by interviewing a particularly hard-hit family in their Long Island summer home.

Just weeks into their vacation, the family's nine-year-old daughter fell seriously ill. All told, six out of eleven members of the household, both family and staff, fell sick.

Soper interviewed the family and discovered that the cook—Mary Mallon—had returned to New York City soon after, apparently healthy as a horse.

New York was in the midst of a typhoid fever epidemic. Typhoid fever is caused by *Salmonella* bacteria and, in 1907, was fatal in 10 percent of infections. Immunization against typhoid fever was four years away and an effective antibiotic was almost forty years away. Carried in feces and usually found in areas with poor sanitation, typhoid fever's appearance in wealthy areas was a mystery.

But Soper was onto something. He questioned the family about the menus prepared by Mary, and discovered that shortly before illness broke out, Mary had served her Sunday specialty: homemade ice cream and fresh peaches. Compared to her hot cooked meals, Soper concluded that "no better way could be found for a cook to cleanse her hands of microbes and infect a family."[1]

Soper searched New York for four months, looking for Mary Mallon. In his search, he identified eight families for whom Mary had cooked, going back several years. Seven families had outbreaks, with twenty-two sick and several deaths.

In a few years, newspapers would dub Mary Mallon "Typhoid Mary."

It took police officers, an ambulance, a female doctor, and a foot-chase over the backyard fence to finally bring Mary in for testing. Mary tried her best to resist, holding her stool sample for as long as she could. Ultimately, nature called, and Mary tested positive for high levels of *Salmonella*. Completely and absolutely healthy, Mary denied ever being sick with typhoid.

She was quarantined on a hospital island for two years, producing typhoid-positive samples the entire time.

Mary sued the health department and lost but was finally freed from the hospital in 1910. She agreed to check in regularly with the health department, and importantly, she agreed never to work as a cook again.

After a while, Mary stopped checking in and, feeling fit as a fiddle, returned to cooking: for a hotel, a Broadway restaurant, a spa, and a boarding house.

In 1915, typhoid fever broke out in a New York maternity hospital. At least twenty-five nurses, doctors, and staff were infected, and two died. The cook was a woman named "Mary Brown" who, surprise, surprise, turned out to be Mary Mallon.

She was again transferred to the quarantine island, where she lived until her death in 1932, adamant that she had never been sick with typhoid.

Mary was the first known case of a healthy carrier in the United States. She was traced to the infection of at least 122 people, including five deaths. In 1907 alone, three thousand New Yorkers were infected with typhoid, most likely due to Mary.

Mary's legacy as an asymptomatic carrier of a deadly disease has informed every widespread disease outbreak since.

Proud Mary

Mary Mallon was quarantined against her will for twenty-five years. She was poor, an immigrant with limited education, and she likely never understood the nuances of being an asymptomatic carrier.

Mary, however, didn't take it lying down. She wrote letters—lots of letters—letters to editors of newspapers, letters to the Board of Health, letters to anyone she thought might listen. She obviously did not help her case when she secretly went back to cooking, but Mary was mad, and understandably so.[2]

The case of Mary Mallon is a difficult one, a case which subsequently changed how we deal with public health crises. Without a doubt, Mary's rights were violated. Health officials never fully explained the situation to her, they made her a pariah, and they isolated her against her will.

Mary's case was botched, yet we understand the need to balance individual rights and the welfare of the community.

Societies cede to governments the responsibilities for public safety, and no public health crisis in history was without controversy. There will always be contrarians who resist safety measures.

Historically, however, Americans have come together for the community good in support of public safety. After years of outbreaks, shutdowns, and quarantines, Americans were overwhelmed with civic pride when the polio vaccine was announced. Church bells rang, factory whistles blew, and people ran into the streets weeping.

All that changed with the COVID pandemic.

Americans had not faced a pandemic since the flu outbreaks of the early twentieth century. The COVID pandemic broke out in a time far more technical, more fast-paced, and with far more expectations of instant solutions.

Public health officials, facing projections of mass hospitalizations (which turned out to be correct), initiated shutdowns of schools, churches, and businesses in an attempt to spread out hospitalizations—we were asked to help "flatten the curve." Orders for masking, social distancing, and crowd-gathering limits went into place.

Almost immediately, we saw intense pushback against the safety measures. And leading the charge? Evangelical Christians.

Evangelical resistance to COVID safety measures, and in many cases outright refusal to comply, was couched in terms of individual rights and freedoms.

We even baptized our resistance as religious freedom.

Rights, Freedoms, and Equal Time

Instead of a conversation about science and the best evidence and research, we had a mutiny about rights. It was not, however, the first time evangelical science denial manifested as a stand for rights and freedoms.

For decades, evangelicals fought evolution under the banner of rights and freedoms.

The trial that started it all, of course, was the Scopes monkey trial. Tennessee had recently passed the Butler Act, making it illegal to teach evolution in public schools. Well, at least not the entirety of evolution—the Butler Act only prohibited teaching the evolution of humans.

The ACLU sent out a press release offering to challenge the law. High school substitute teacher John Scopes agreed to be the guinea pig (read: sacrificial lamb), admitting to the offense of teaching evolution.

The trial folderol lasted days, yet no scientific evidence or testimony was included. The jury deliberated only nine minutes, convicting Scopes of teaching evolution.

Forty years later, the United States Supreme Court ruled that laws forbidding the teaching of evolution are unconstitutional.

Undaunted, creationists campaigned instead for equal time in classrooms, with "creation science" taught alongside evolution. The ICR encouraged citizens to lobby their local school boards to include "creation science" in the biology curriculum. By the early 1980s, "equal time" laws had been introduced in twenty-seven states. Most were defeated, but five states passed laws requiring equal treatment for creationism.

The Louisiana equal time law was challenged in the Supreme Court in 1986 in *Edwards v. Aguillard*. A similar Arkansas law had been struck down at the state level a few years before. Supporters of the Louisiana equal time law avoided the creationist arguments used in the defeated Arkansas case and took a different approach.

This time around, the game plan was protection of rights and freedoms.

Even so, the Supreme Court ruled that creation science was a religious doctrine and struck down the Louisiana law. The Supreme Court ruling, however, did not stop a new crop of rights and freedoms cases in district courts and appeals courts across the country:

- In Illinois, a teacher argued that his right of free speech was violated when he was prohibited from teaching creationism.
- In California, a teacher argued that her rights were violated when she was required to teach evolution in biology class.
- In Minnesota, a teacher argued that he had the right to teach both evolution and creationism.
- And in Georgia, a school board fought for the right to place a sticker in all biology textbooks, stating "Evolution is a theory, not a fact, regarding the origin of living things." (Penalty flag for misrepresenting what science means by the term *theory*.)[3]

Eighty years after the original, in a case described as the Scopes trial of the twenty-first century, advocates of the intelligent design variant of creationism argued that the Constitution protects the right to learn about intelligent design, and to deny that right is an assault on freedom.

The federal judge in *Kitzmiller v. Dover* (2005) disagreed and ruled that intelligent design is a religious construct and as such, cannot be taught in public schools.

Defeated (for now) in the courts, the fight for the freedom to teach against evolution continues in local and state boards of education. The fight is usually in the form of a curricular requirement to teach the "controversy" of scientific theories, without specifically mentioning evolution. Undeniably, the theory in mind is evolution—nobody wants to debate gravity or molecules.

Geared Up and Ready to Go

They couldn't have known it, but evangelicals were gearing up for a new fight.

Going into the 2016 presidential election, evangelicals overwhelmingly believed that religious intolerance was on the increase in the United States.[4] "Position on religious liberty" was the number one reason given by evangelicals for deciding who got their vote for president.[5]

It didn't take long for the COVID pandemic to turn political, and when it did, it opened a new front in the fight for religious liberty.

Shutdowns presented a complex set of challenges. One fix exposed other problems: close schools and kids disappear. Close restaurants and businesses suffer. Cancel events and chaos ensues.

Deborah Birx, the White House coronavirus coordinator, along with Anthony Fauci, drafted guidelines for careful and safe reopenings. A documented decline in cases was crucial to the reopening plan. Birx and Fauci clearly articulated their plan to the public, or so they thought.

Just weeks into the shutdowns, President Trump called for "liberation." In all-caps tweets, Trump called on his supporters to "LIBERATE VIRGINIA!" and "LIBERATE MICHIGAN!"[6]

The message was loud and clear and had the bully pulpit of the president. We need to be liberated, and people who need liberation are people whose rights and freedoms are under assault.

The cry for rights and freedom emerged from Washington and spread like fire across the states. In some cases, literally.

Ashley Smith, creator of the viral video "Burn Your Mask Challenge," urged others to do the same, using the hashtag #IgniteFreedom. According to Smith, "masking, mandatory temperature checking, things like that, these are just ways our freedoms are being eroded."[7]

The award for the most cringey video, however, goes to Washington state representative Jim Walsh. Walsh appeared in a livestreamed speaking event wearing a yellow Star of David. He did it to prove his point: "Denying people their rights can lead to terrible outcomes." Sadly, he was not the only lawmaker to equate COVID restrictions with the Holocaust.[8]

It is no secret that self-identified conservatives, Republicans, and supporters of President Trump disproportionately resisted COVID restrictions. It is also no secret that evangelicals are disproportionately conservatives, Republicans, and supporters of President Trump.

"Freedom is under assault!" was the cry of the president and his allies. Why, then, would evangelicals listen to Fauci or Birx or any of the epidemiologists tracking the pandemic? After all, Fauci, Birx, and their ilk were the ones curtailing our rights.

Early on, the president drew a line in the sand: the virus is overblown, and restrictions are a violation of rights. Evangelicals crossed the line and stood with him. Resistance to restrictions was a badge of courage and a holy war.

Curtis Chang is a theologian and author. Chang was one of a few evangelical voices in the wilderness, calling evangelicals to listen to the science, comply with restrictions, and take the vaccine. Chang produced a set of videos encouraging Christians to "think biblically" about all things COVID.

Thinking biblically was easier said than done, however. As Chang observed, "Once Trump set the Republican culture down this path, he made it very difficult for evangelical leadership to lead."[9]

Robert Jeffress is the pastor of the fifteen thousand-member First Baptist Church in Dallas. He is a regular guest on various Fox News shows and was a spiritual advisor to President Trump.

Echoing the apostle Peter's ultimatum when the government tried to shut *him* down, Jeffress declared, "We need to obey God rather than man." Jeffress praised a fellow pastor who instructed his congregants to circumvent government regulations by holding "a funeral for a turkey" instead of a Thanksgiving celebration.[10]

The fight was on. We baptized political obstinance as religious freedom and off we went.

It wasn't long before evangelicals adopted a popular political slogan and made it their own. In the muted-colors art of a classic children's Bible story book, we see Jochebed and Miriam, knee deep in the river, putting chubby baby

Moses into a wicker basket. The caption reads: "MOSES's MOTHER DID NOT COMPLY."

I admit to a giant eyeroll when I saw that one in my Facebook newsfeed.

Becky. It's a mask. No one is trying to throw Junior into the Nile.

The Western Front

The front line in the fight for religious liberty quickly moved to the San Fernando Valley.

Grace Community Church in Los Angeles had a prepandemic weekly attendance of 8,000 in a sanctuary seating 3,500. John MacArthur, pastor of Grace since 1969, is regarded as a Christian elder statesman by evangelicals.

In September 2020, Fox News host Laura Ingraham introduced a segment with MacArthur by referencing "petty tyrants" in state, county, and city governments, as well as in the health department, all of whom are launching an "assault on your inalienable rights, like your right to religious liberty."[11]

With multiple references to the Constitution and the backing of the president, MacArthur explained his decision to defy California's pandemic orders. MacArthur's closing comment was concise: "Bring it on."[12]

Grace Community Church became the model, the exemplar, the mother church for other churches defying court orders and fighting the good fight for religious freedom. In October 2020, MacArthur posted a video on Twitter, challenging evangelicals in all caps: "OPEN YOUR CHURCH."[13]

Claiming targeted discrimination by the health department, MacArthur peppered the just-over-three-minute video with references to the Constitution. The Constitution, according to MacArthur, gives the church the right to assemble as they please.

Owen Strachan called the video a "nitrous burst" speaking with "pastoral authority." Strachan mocked calls for consideration of public health, saying, "What Marxism tried to do for decades, with not great success in America at least, 'public health' did in a few weeks."[14]

So, there you go. Not only are evangelicals being denied their constitutional religious rights, but they are also living under full-blown Marxism.

Winnowing the Weak

Located halfway between London's posh Barbican theater district and the famed Smithfield Market is a plot of land with a gruesome past. Excavation in the 1980s revealed bodies—loads and loads of bodies. So far, six hundred bodies have been cataloged, but there are probably more than two thousand.[15]

The year was 1348, and after devastating the continent, the Black Death had arrived in England. By 1350, one-third of Britain was dead of plague.

London's churchyards could not accommodate such a colossal demand, so the city created a five-acre burial pit known as East Smithfield. At the height of the plague in London, two hundred people per day were buried in the mass grave. Hastily, the dead were buried, some neatly lined up, others tossed in haphazardly.

The plague struck hard and fast in Britain and across Europe, and in a short time it overwhelmed the population.

Prevention measures (as best as possible in the prescientific, pre-modern-epidemiology days of the fourteenth century) were practiced. Avoiding the "bad air" thought to spread plague meant shuttering doors and windows, and suspending access to family, friends, and neighbors. Households with plague were quarantined.

But for the survivors, the story took a bit of an uptick. For generations after the plague years, survivors were healthier and lived longer than did the general pre-plague population. Interestingly, modern genetic studies suggest that some survivors had innate resistance to the plague or to its fatality.

However, there was probably an additional environmental factor. As much as half the population died in some areas. With the weak winnowed out, survivors had access to more food, more meat, more and better bread.

After years of lockdowns and quarantines, Londoners were ready to return to normalcy. Blinking in the plague-free sunlight, survivors emerged, eager to get on with life, with socialization, and with commerce.

Just months into COVID lockdowns and with no end in sight, many Americans were anxious for a quick fix. Hydroxychloroquine and America's Frontline Doctors promised a cure. The *Plandemic* film told us that fear was the *real* virus.[16]

The Great Barrington Declaration promised us the solution we longed for. Drafted by three public health scientists from prestigious universities, the document was released by the American Institute for Economic Research, a libertarian think tank in Massachusetts.[17]

The Declaration argued that COVID should be allowed to spread uncontrolled among the healthy, while presumably protecting the vulnerable. The result of such a strategy (according to the document) would be "herd immunity" without the use of a vaccine.

Despite the résumés of the Declaration authors, the document did not include any references to studies or scientific evidence of any sort. The Declaration was denounced worldwide as a dangerous fallacy unsupported by evidence, ethically problematic, and "total nonsense."[18]

In cases of viral diseases for which we have a vaccine (like measles), it's true: we rely on herd immunity to protect the very young, the immunocompromised, and the few who (unknowingly) do not mount an immune response following a vaccine. But in the case of measles, herd immunity is not achieved by allowing a deadly disease to sweep through an entire population, killing and maiming many but leaving a few survivors with resistance. With measles, herd immunity is achieved by vaccination without the devastating effects of the actual disease.

When the Great Barrington Declaration was published, no COVID vaccines were available. Any herd immunity would be achieved by massive outbreaks of disease.

The Declaration proposed that we allow COVID to run rampant in low-risk populations while isolating the vulnerable. Although the young and healthy are at lower risk (statistically) from death, they are not exempt from serious disease or the impact of long COVID. The Declaration made no mention of overwhelmed medical facilities from COVID gone wild.

The Great Barrington Declaration was hailed by politicians as a way to return to normalcy. It was trumpeted from the White House as the way to immediately reopen the economy.

Evangelicals flying the flags of noncompliance were all on board. And to those chafing under the assault on their religious right to assemble, unmasked, in large groups, the Declaration was a godsend.[19]

The Great Barrington Declaration was clearly scientifically problematic, but its embrace by Christians is disconcerting. Theoretically, we would protect those at highest risk, presumably isolating the sick and elderly in their homes or in residential care.

Maybe we should talk to the Zubia family. In October 2020, the very month the Great Barrington Declaration was published, the four Zubia siblings contracted COVID. Their father tested negative. All were employed—at a grocery store, in plumbing, as a mechanic—not the kinds of jobs where you can work from home.[20]

The siblings and their father lived together in a tiny one-bedroom, one-bathroom apartment in Los Angeles. The sisters shared a twin bed. It was all they could afford, but they made it work.

When the siblings contracted COVID in October, they were distraught. Their fifty-nine-year-old father, Jose, had diabetes.

The family wore masks—even slept in them. They pulled the covers over their heads. They sprayed the bathroom with disinfectant. They kept the door open.

Predictably, Jose contracted COVID and was hospitalized. The day he was put on a ventilator, Joanna, his daughter, was admitted to the intensive care unit (ICU).

As Jose was dying, only one sibling was allowed to visit.

Before Jose died, the ICU doctor visited Joanna in her hospital room. The doctor asked if the family had been socially distancing.

Joanna began to cry. The ICU was first time in her life she had her own room.

"That's when you realize social distancing and working from home are privileges not everyone has," the doctor said.[21]

During the pandemic, Americans in poorer counties died at twice the rate of those in wealthier counties.[22]

Advocating for a winner-take-all, survival-of-the-fittest, let-the-virus-roar-through-the-population approach to the pandemic disregards the people who can't work from home. It disregards people who live in multigenerational homes and those who share homes with at-risk family members. It disregards people with tenuous incomes and access to health care.

It was not without reason that the Great Barrington Declaration was called "ethically problematic" by the head of the WHO.

Winnowing the weakest among our community is *ethically* problematic, yes, but it is also love-your-neighbor-as-yourself problematic. It is whatever-you-did-for-the-least-of-these problematic.

What Determines Our Rights?

The COVID pandemic was not the first public health crisis faced by Christians.

Christians built the first public hospitals. Care for the sick was not unknown, but Christians were the first to provide care for those outside their kinship group. Ancient historians noted that in times of plague, Christians cared for all, Christian or not.[23]

In the first century, they knew we were Christians by our love.

In the twenty-first century, they'll know we are Christians by our fight for rights.

Tim Diebel is a retired pastor and a religion writer. Early in the pandemic, Diebel quoted his daughter in this convicting observation: "I wish the CDC hadn't said that masks primarily protected other people, rather than the wearer. I wish they had lied and told us that masks were the simplest way to protect ourselves. We don't seem to care about protecting others; only ourselves."[24]

In the first century, they knew we were Christians by our love. In the twenty-first century, they'll know we are Christians by our fight for rights.

The COVID pandemic exposed an evangelical fervor for personal rights and freedoms. Birthed in the fight against evolution in schools, the fight for rights took on an overtly political mood during the pandemic. And what do you get when you mix religion with politics?

You get politics.

In a time of crisis, in a culture of rugged individualism, in the land of the free, what determines our actions? The law of love, or the rights guaranteed in the Constitution?

9

Anthony Fauci Hates Puppies

The hospital ward has beds for fifty children.

Dr. Banting and his associate quietly step inside the silent room. Bed after bed is occupied by a child, comatose and dying from diabetic ketoacidosis. Grieving parents sit bedside, awaiting the inevitable.

The year is 1922. Frederick Banting and his assistant Charles Best had been studying type 1 diabetes for the past two years.

Marjorie was part of the team, also. Just weeks before, Marjorie posed for a photo with Banting and Best on the tar paper and gravel roof of the laboratory. Best was smiling, his tie flapping in the breeze, and Banting, wearing his lab coat, was leaning in toward Marjorie.

Marjorie, however, was pensive, maybe even a bit wary. Her coat was a soft brown and white. Marjorie's ears, however, were flattened on her head and her tail was between her legs. Marjorie, the pretty little stray collie, was about to make history.

There were others before Marjorie, thirty-two, in fact. But Marjorie was the lone survivor.

Banting and Best induced diabetes in the dogs by removing their pancreases. Banting and Best then extracted insulin from the dog pancreases and injected it back into the now-diabetic dogs. This was uncharted territory: How much insulin was needed to keep a diabetic dog alive?

Marjorie lived seventy days. Marjorie did not die from diabetes, however, but from a postsurgical infection.

We've known about type 1 diabetes for centuries. The condition is well documented in ancient Egyptian, Hindu, and Chinese medical records. Aretaeus of Cappadocia, a second-century Greek physician, described diabetes as "the melting down of flesh and limbs into urine."[1]

When the pancreas no longer produces insulin, people with type 1 diabetes cannot metabolize carbohydrates. Without insulin, patients waste away and suffer greatly, no matter how much they eat.

The death rate for type 1 diabetes, prior to insulin therapy, was 100 percent.

By the nineteenth century, we understood the role of carbohydrate metabolism in diabetes. Doctors attempted to control the disease with diet—prescribing only meat or fat for suffering children. On severely calorie-restricted diets, children lost weight and were critically malnourished.

Still, instead of death in a few weeks, children might survive a year.

Encouraged by their success with dogs in the lab, Banting and Best injected an emaciated fourteen-year-old boy with cow insulin in January 1922. Daily injections over the next twelve days dropped the boy's blood sugar until it was no longer detectable in his urine.

Soon after, Banting and Best walked into a solemn and quiet ward in a Toronto hospital filled with comatose children. Banting and Best moved from bed to bed, injecting each child with insulin.

As they injected the last child . . . the first child woke up.

One by one, all the children awoke from coma. A room thick with impending death became a place of hope.

The 1923 Nobel Prize in Medicine was awarded to Frederick Banting and John Macleod, in whose laboratory Banting and Best worked. Banting shared his prize with Best.

The men won the prize, but don't forget the pups—heroes all.

Marjorie and the other dogs received loving care from Banting and Best. By all reports, Banting and Best considered the dogs friends and coworkers.[2]

It is easy from our perch in the twenty-first century with all its biotechnology to shake a disapproving head at animals used in medical research. I'm not going to launch a debate here about animals in the lab, but I want to acknowledge the Marjories in the past and all the heroic present-day Marjories with gratitude.

Even in the twenty-first century, animals are often the best tests for preventions and treatments of deadly and disabling human illnesses. Most of us accept this and simultaneously advocate for humane treatment of research animals. Most of us are grateful for the sacrifices of the four-legged heroes who allow us and our families to live healthy lives.

The Villain

The photo was shocking: two obviously sedated beagle puppies, heads locked in mesh cages, screamed from the headlines. Inside the mesh cages were hundreds of sand flies, feasting at will on the helpless, drugged-up pups.

What heinous Gotham City villain was responsible for such cruelty?

In the fall of 2021, a conservative watch dog group called the White Coat Waste Project broke the story, published the photos, and identified the villain.

The culprit responsible for this reprehensible act was none other than the director of the National Institute of Allergy and Infectious Disease and the public face of America's response to the COVID pandemic: Dr. Anthony Fauci.

Despite the provocative headlines, the National Institute of Allergy and Infectious Disease (NIAID) did not authorize or fund this study. The watchdog group incorrectly associated the puppies-in-a mesh-cage study (a Tunisian study) with Fauci and the NIAID.

The NIAID did fund a study designed to test a vaccine for leishmaniasis, a parasitic disease transmitted by sand flies. Dogs were vaccinated and released into an enclosed open space designed to mimic natural exposure to the flies. No pups were drugged or fed to flies.

But no matter. The narrative was set.

Before you know it, "Fauci" and "puppies" were trending, with hundreds of thousands of searches. Donald Trump Jr. sold "Fauci Kills Puppies" hoodies online.

Politicians already critical of Fauci's handling of the pandemic and (apparently) recent converts to the animal rights movement weighed in. Senator Ted Cruz (R-TX) tweeted that Fauci was "torturing puppies."[3] Florida Governor Ron DeSantis called Fauci a "mad scientist."[4] And they weren't the only ones.

Clearly, Dr. Fauci is not only an overpaid and underqualified bureaucrat, a liar, and the biggest COVID-flip-flopper on the planet, he's worse.

Anthony Fauci hates puppies.

From the outset of the pandemic, no stereotype was left unturned—mad scientist, super villain, charlatan. When

Dr. Fauci advised against large holiday gatherings, he was the Grinch who stole Christmas.

Dr. Seuss was hauled yet again into the scrum, months before Puppygate. When it was announced that six Dr. Seuss books would no longer be published, outcry was strong in evangelical quarters. Angst over the vanquished books soon gave way, however, to a runaway meme populating social media posts and screen-printed on countless T-shirts: "I trust Dr. Seuss more than I trust Dr. Fauci."

Medical, Not Political

Long before he was the face of the COVID fight, Anthony Fauci had your back.

When it was an unknown, scary, "happens to other people" disease, Fauci brought AIDS research and treatment to the forefront and brought scared and frustrated AIDS sufferers to the table. What was at one time a death sentence is now managed by pills.

He's been advisor to and beloved by presidents—Reagan through Obama—from both sides of the aisle. Ebola, measles, swine flu, bird flu—everyone wanted his input. He was awarded the Presidential Medal of Freedom.

Dr. Fauci stands up for science, even when it's politically unpopular, and he has for fifty years: "What I learned is don't be political. Be medical. Be scientific. . . . It's when you get into the politics that you get in trouble."[5]

Margaret Sullivan's *Washington Post* headline says it all: "Only in Our Anti-Truth Hellscape Could Anthony Fauci Become a Supervillain."[6]

How did a fifty-years-behind-the-scenes research scientist become the most lampooned man in America? And why were evangelicals front and center, wielding the lampoon?

Red Flags

How did evangelicals manage to discredit Anthony Fauci and broadly discredit the majority of pandemic scientists?

A popular Christian curriculum widely used in homeschools and private schools introduces us to two scientists on the very first page of the earth science textbook. Both are white. Both are male. Both wear glasses, beards, and sensible long-sleeved polos. Both scientists pop up throughout the book, offering commentary on the topic at hand.

This is where the similarities end. We are told that these two scientists, while both highly educated, often disagree regarding matters of science. Although the two examine the same evidence—the same rock, the same fossil, the same DNA—one is right, and one is wrong.

Which of the two, then, is the trustworthy guide? Let's ask each the same question—you be the judge: How did the Grand Canyon form? The scientist in the green polo talks about sediments, shallow seas, and erosion over millions of years. The scientist in the red polo talks about sin, judgment, and a global flood—all in under seven thousand years. Same canyon, wildly contradictory answers. Both cannot be correct.

To decide which scientist to believe, we need one more bit of information: the scientist in the green polo is a *secular* scientist. Hoist the red flag! If there is a single term consistent across all creationist literature and events, it is the term *secular*. When it comes to scientists, there are two, and only two, categories: (1) secular scientists and (2) godly, Bible-believing scientists, also referred to as *creation scientists*.

The term *secular*, when paired with the term *science* or *scientists*, is poison—a red-flag warning "this person is not to be trusted" or "this information is not to be believed."

Our Christian textbook tells us that the red-polo scientist is a trustworthy scientist because he holds the Bible as the only reliable source of truth for science. Red-polo scientist is certain the world is young because the Bible teaches it is.[7]

It is important to note that the creationist definition of a secular scientist does not exclude people who identify as Christian. On the contrary, many people of faith have no problem accepting the scientific evidence for evolution and an old earth.

Creationist organizations, however, describe a secular scientist as one who refuses to view all scientific evidence through a biblical lens. In other words, a scientist (be they Catholic, Protestant, Orthodox, or even evangelical) who does not interpret evidence through the filter of a literal Genesis is accepting man's faulty and atheistic reasoning over God's firsthand witness of creation—the mark of a secular scientist.

Secular science is described in creationist publications as popular ideas attractive to scientists who want to be accepted by the mainstream scientific community. It's not that creation scientists don't know the secular science; they are simply smart enough to reject it.[8]

The secular red flag is not planted only by organizations dedicated to creationism. Albert Mohler, long-time president of the Southern Baptist Theological Seminary, in an interview with Kurt Wise, discussed the bankruptcy of belief in evolution and an old earth. Their conclusion? Secular scientists block creationism in their secular universities because they don't want to be accountable to God.[9]

Furthermore, secular scientists cannot be trusted to employ the long-respected standard of scientific truth-seeking, the scientific method. The problem, according to John Baumgardner of the ICR, is not the method but the scientists. The scientific method fails, says Baumgardner, because of those who are attempting to apply it.[10]

In a remote, almost inaccessible cave system in South Africa, thousands of fossil bones belonging to a never-before-identified human ancestor were discovered. A team of cavers and divers assembled by paleoanthropologist Lee Berger retrieved at least fifteen individuals from deep within the Rising Star cave system. The discovery of a new human ancestor, *Homo naledi*, was exciting for many reasons and holds a wealth of new information about human evolution.

When the discovery of *Homo naledi* was announced in 2017, the ICR ran the story, including quality photos of *H. naledi* skulls. The article focused on a "cherry-picked age" for the fossils and linked the reader to an extensive ICR resource. The title of the resource? "Reasons to Doubt Secular Ages."[11]

If you want to discredit science, simply hoist the secular red flag.

Science Denial Lite

The messaging that science and scientists are not to be trusted ranges from subtle to flagrant.

On the subtle end of the scale is the 2017 film *Is Genesis History?*, a pretty and pleasant film from the popular evangelical organization Focus on the Family. The film was widely released in theaters and is still available for purchase and streaming.

Is Genesis History? is beautifully filmed and well made. The film focuses on interviews with theologians, philosophers, and many creation scientists. Those interviewed all seem knowledgeable, and all seem really nice. A sequel is planned for release in 2022 or 2023.

The antiscience message of the film is fairly subtle—here's what secular science says, but it's wrong—wrong about geology, about biology, about astronomy. They're very nice about it, though.

But the message is unmistakable: the overwhelming majority of working research scientists, worldwide, are wrong.

To understand the impact of *Is Genesis History?*, let's put it in another context. How would you react to a film claiming most scientists are wrong about germ theory? How would you react to a film claiming most scientists are wrong about gravitation theory? Evolution theory has the same degree of support among biologists.

Subtle accusations like those in *Is Genesis History?* take a subliminal toll on the legitimacy of scientists, but for decades, evangelicals have been overtly told that scientists ignore and dismiss evidence for special creation and a young earth.

Worse than that—secular scientists actually *fabricate* evidence for evolution.

Forgeries and Fakes

Englishmen love their cricket, and apparently, they have for a million years.

The year was 1912. Amateur archaeologist Charles Dawson announced to the world the discovery of the "first Englishman." Near the town of Piltdown in England, Dawson found fossils of teeth and a human-like skull. A jaw was also found, but it was more ape-like. Several tools were found nearby, including a sculpted elephant bone that looked for all the world like a cricket bat.

Dawson's find was heralded as a million-year-old apeman, an exciting missing link in human evolution. Take that, you French and Germans with your fancy Cro-Magnon discoveries! We have the first Englishman, and he's half ape!

Dawson died in 1916, and in subsequent decades things just didn't pass the smell test with Piltdown man. Anthropologists were making more early human fossil discoveries and these had little in common with Piltdown man. In 1949, more tests and more studies were conducted, and it turns out Piltdown man was a fraud. A cobbled-together, glued-together, doctored-up fraud.

From our vantage point in the twenty-first century, Piltdown man is just an interesting story and a blip on the screen compared to the wealth of fossil and genetic evidence we have for human evolution. Yet, Piltdown man is trotted out and used time and time and time again by creationists as the go-to example of how fossils are made-up frauds perpetuated by scientists.

Ignored is the fact that it was actually secular scientists who exposed the fraud of Piltdown man more than seventy years ago.

And then there's the Piltdown chicken.

In 1996, a poor farmer in the Liaoning province of China stumbled upon a small, turkey-sized dinosaur. The little dino belonged to the theropod group of dinosaurs, a branch it shares with *Tyrannosaurus rex*. But it wasn't the dino's pocket-pet dimensions that stunned paleontologists.

It was the feathers.

Sinosauropteryx ("the China dragon bird") was the first of an absolute treasure trove of feathered dinosaurs discovered over the next two decades, primarily in China. Finds included the duck-sized *Caihong juji* with its hummingbird-colored feathers and the one-and-a-half-ton *T. rex* cousin, *Yutyrannus huali*.

But for the moment, let's go back to the 1990s, just three years after the discovery of *Sinosauropteryx*. Same province in China, different farmer.

The discovery of *Sinosauropteryx* triggered a fossil gold rush in China. Farmers pored over their fields, hoping to find the next exciting feathered dinosaur fossil and sell it for tens of thousands of dollars to a collector or a museum. When you make only a few dollars a year, a feathered fossil is paydirt.

In 1996, a fascinating new find was announced. Called *Archaeoraptor*, it was hailed as a marvelous missing link in bird evolution, with the arms of a primitive bird and the tail of a dinosaur.

Against the advice of paleontologists, *National Geographic* broke the story of *Archaeoraptor*. Paleontologists cited many problems with the fossil and the lack of peer-reviewed studies. You guessed it—*Archaeoraptor* was a jerry-rigged fake, a forgery of bird and dinosaur parts. The press dubbed it the "Piltdown chicken."

Archaeoraptor was not a scam perpetrated by scientists but a fraud created by a desperately poor farmer hoping to make a buck.

Creationist publications had a heyday: "This wouldn't be the first time that *National Geographic*, in its eagerness to proselytize for the evolutionary faith, has rushed into print with 'evidence' that has turned out to be a hoax or an overblown claim that was later discredited."[12]

But decades later and despite hundreds of feathered dinosaur fossils and well-established dinosaur-to-bird-evolution evidence, *Archaeoraptor* is still the poster boy for bird-evolution denial. Here's a 2016 posting from Answers in Genesis: "In three months, over 100,000 young people saw the 'proof' for dinosaur-bird evolution on display at

National Geographic's headquarters in Washington, D.C. It was all FAKE. The supposed fossil was fake. The artwork and article in *National Geographic* described a fake. What influenced so many students touring the *National Geographic* exhibit in Washington was the display of a fake."[13]

Like Piltdown man, *Archaeoraptor* is a blip on the screen, an interesting story in a wealth of bird evolution evidence, and a cautionary tale in peer review.

Unfortunately, what most people who reject evolution know about evolution comes from anti-evolution apologetic sources. If this is how you learned the "facts" of evolution science, it would be easy to think all human fossils were frauds. You could be excused for thinking that dinosaur-to-bird evolution was wishful thinking.

And without a doubt, you'd think secular scientists perpetuated these frauds.

Vilification of Scientists

J. W. Wartick is a religion blogger who writes extensively about the evolution of his Christian faith. Like lots of kids, Wartick was obsessed with dinosaurs as a child. His parents regularly drove him to the local library where he checked out every available dinosaur book in the children's section. He memorized all the facts—what the dinosaurs ate, where they lived, what they looked like—with one exception: "The dates, I was told, were wrong. Whenever I saw 'millions,' I was told to ignore it. . . . I learned to just run my eyes over any time it said 'millions of years,' because that was wrong. . . . I didn't realize how odd it was that the books and scientists seemed to be right about, say the ecosystem the dinosaurs lived in, or their diets, while simultaneously being totally wrong and even untruthful about how long ago they lived."[14]

Untruthful? Wartick continues, "I used the word 'untruthful' because part of what I was taught, whether directly or through creationist literature, was that scientists weren't just *wrong* about the age of the Earth or when dinosaurs lived. . . . They were actively *lying* about it."

On the flip side, creationist Gary Parker also felt misled as a child: "Like most Americans, I was mis-taught in grade school that it takes millions of years and tremendous heat and pressure to turn sediments (like sand, lime, or clay) into rock (like sandstone, limestone, or shale). We all know better. . . . The concept of evolution touted in textbooks, then, is based on phantoms and figments of the imagination."[15]

The Dallas-based Institute for Creation Research (ICR) has new and impressive facilities with an education center and museum. ICR is all set up for the classic school field trip—a science museum, planetarium show, live presentations by ICR scientists, and of course, a gift shop. And field trips abound! A steady stream of homeschoolers, Christian schools, and private groups visit ICR's Discovery Center where they learn that secular scientists practice "blind guesswork" with evolution driving the bus.[16] By contrast, secular museums waste taxpayer dollars with their "deceptive" displays.[17]

Some of the most egregious denigrations of secular scientists can be found in popular homeschool and Christian private school curricula. Thumb through the earth science and biology textbooks of Abeka, Bob Jones University Press, and Accelerated Christian Education and you'll find that secular scientists guess, distort the truth, make mistakes, ignore facts, have no proof, are motivated by politics, and are blinded by their atheistic worldviews.

The Creation Science Research Center in San Diego, California, creates resource materials for homeschoolers. In the introductory materials, we learn that secular scientists (as well as secular science educators) teach a "distorted" science: "Those who today control science and education have adopted a false definition of science."[18] With such an introduction, why would a homeschooler trust anything a secular scientist says?

Ask the questions. Question the answers. That's fair, but that's a far cry from broadly painting secular scientists with the brush of dishonesty.

Words matter. Characterizations matter. When you read or listen to creationist presentations, pay attention to the words used to describe secular science and secular scientists. Is the language peppered with descriptors like "allegedly," "spin," and "dubious"? Does the speaker affect a snarky tone when describing the evidence accepted by secular scientists?

Ask the questions. Question the answers. That's fair, but that's a far cry from broadly painting secular scientists with the brush of dishonesty. How we characterize the messenger determines how we receive the message.

Words Matter

During the pandemic, scientists were accused of having a political agenda. Scientists with decades of public health service were called charlatans, liars, and worse by evangelicals.

Politicians and pundits with strong evangelical support led the charge from their prime-time bully pulpits. Media-savvy pastors chimed in.

When he retired in 2022, Francis Collins was the longest-serving head of the NIH. Before that, he led the history-making Human Genome Project. A world-class geneticist, Collins identified the genes responsible for cystic fibrosis and Huntington's disease. Collins received the Templeton Prize, the National Medal of Science, and the Presidential Medal of Freedom.

Francis Collins is also an unapologetic evangelical Christian. With his résumé, you'd think Collins would be the Tim Tebow of scientists.

Yet Collins is not celebrated in all evangelical quarters. When Collins retired from the NIH, the Discovery Institute's John G. West wrote that grief, not celebration, was the appropriate evangelical response: "Collins served his purpose at NIH by providing cover for the secularists to do what they wanted to do anyway."[19]

Despite a decades-long reputation as someone who successfully navigates government without compromising his science, Collins is regularly accused of being a bureaucrat and politician. The personal attacks are worse—Collins admits to being troubled by the loads of vitriolic emails he receives after each appearance on Fox News.[20]

But the worst evangelicals had to offer was reserved for Anthony Fauci. Like Collins, Fauci has decades of service in research and health administration. And like Collins, Fauci has a reputation for steering clear of politics and sticking to the science. But when Fauci declared on *Face the Nation* that he represents science, he was thoroughly excoriated by Republican senators for the *audacity* of the claim. "The absolute hubris of someone claiming THEY represent science," said Senator Rand Paul. "It's astounding and alarming that a bureaucrat would even think to claim such a thing."[21]

And Fauci is not only a bureaucrat, he's dishonest, political, and partisan. He's Mussolini! No, he's Josef Mengele![22]

Evangelical pastors called Anthony Fauci "Fauxi" and a traitor to America. Pastor Rick Wiles of Florida called Fauci a liar and called for waterboarding.[23] Another pastor called science an idol worshiped by Fauci, the high priest.[24]

Suddenly, "I trust Dr. Seuss more than Dr. Fauci" sounds rather tame.

There were many reasons for evangelical science denial in the pandemic, and the reasons are nuanced and varied.

But it all begins with shooting the messengers.

10

The Earth Is Running a Fever

It's ten a.m. on a windy March morning in 2013. The scene is overwhelmingly brown—dirt road and dust, with nary a tree in sight or a blade of grass.

We see a faraway group of people at the end of the long dirt road. There are about twenty of them—men, women, children, babies in strollers. As they come closer, we hear strains of a cappella singing, interspersed with the huffing and puffing of hiking in the wind.

> Our God is greater, our God is stronger
> God, You are higher than any other
> Our God is Healer, awesome in power
> Our God, our God. . . .[1]

Each Saturday, a group gathers to walk the four-mile road circling the Cargill meat-packing plant in Plainview, Texas. They walk, they sing, they pray for rain.

Just a few weeks before, the plant shut down, laying off 2,300 workers. Overnight, 10 percent of the area's workforce was out of a job.

A three-year drought in West Texas devastated the cattle industry. Ranchers sold off their herds. No cows, no beef. No beef, no need for a huge meat-packing plant.

There are more than seventy churches in Plainview, a town of twenty-two thousand. Of the calamitous plant shutdown, one former Cargill employee declared, "I think it's biblical."[2]

No one thinks humans played a role in the devastating drought. Public opinion says it is all part of a natural cycle, a cycle solely in the hands of God: "There's only one man who knows how much rain we're gonna get and that's God. And he's not a scientist. I'm not putting much faith in what they say," said one Plainview resident.[3]

Climate Science 101

We've known since the mid-nineteenth century that burning coal, gas, and oil blankets the earth with heat-trapping gases.

The first time an American president was warned about the dangers of increasing levels of carbon dioxide in the atmosphere was in 1965. Climate scientists warned President Lyndon B. Johnson about the consequences of human-caused global warming and suggested mechanisms to ward off catastrophe.

President Johnson released the report for publication, but his administration was more focused on visible and immediate environmental issues like water pollution and roadway litter. Texas highways are beautiful thanks to the advocacy of his wife, Lady Bird Johnson. Burning fossil fuels was put on the back burner, so to speak.

The overwhelming consensus—essentially 100 percent—of working, publishing climate scientists agree: global warming is real, and humans are the cause.[4]

The science that tells us the earth is dangerously warm and humans are the cause is the same science we use to engineer our stoves and refrigerators and to fly our airplanes. If someone objects to the science of climate change, it is likely for a reason that is not a scientific reason.

Weather is what you see when you look out the window and what you feel when you walk outside on any given day. Climate is how the weather behaves over a long period of time. The National Oceanic and Atmospheric Administration puts it this way: climate is what you expect; weather is what you get.

Climate looks at long-term trends, over twenty-to-thirty-year spans. Climate scientists aren't concerned with a random wet or dry or cold or hot year.

Climate scientists are concerned because we are having higher highs, *every* year.

When we burn fossil fuels and cut down our forests, we generate greenhouse gases—primarily carbon dioxide. Methane gas, another greenhouse gas, comes from landfills and agricultural practices. When released into the atmosphere, greenhouse gases wrap around the earth like a blanket, trapping the sun's heat and raising global temperatures.[5]

If someone objects to the science of climate change, it is likely for a reason that is not a scientific reason.

And yes—some carbon dioxide in the atmosphere is needed. Normal levels of carbon dioxide are the light blanket the earth needs in order not to be a frozen planet. By burning fossil fuels, humans are adding an extra-large, extra-thick, extra-heavy weighted blanket to the light covering we need.

The earth is over one degree Celsius warmer than it was in the late nineteenth century. By the end of this century, the global average temperature is projected to be over three degrees higher.[6] For every one and a half degrees Celsius of warming, we will have more intense heat waves, longer warm seasons, and shorter cool seasons. With two degrees of warming, we will be at critical thresholds for agriculture and health.[7]

But it's not just about the heat, says this Texan, who is at the moment writing on a patio on a hundred-degree-plus Texas summer day.

Increased global temperatures intensify the water cycle, giving us heavier rainfall and serious flooding in some areas, drought in others. Warmer air holds more moisture, and when a storm rolls in, all that water unloads. Higher temperatures also mean more water evaporation from the soil, strengthening high pressure systems. High pressure systems then push rainstorms away from the area, bringing drought.

Hot, dry conditions intensify and enlarge wildfires in fire-prone areas of the globe.

Increased global temperatures melt the planet's ice sheets, glaciers, sea ice, and seasonal snow cover. As sea levels rise, coastal areas flood and coastlines erode.

Oceans absorb over 90 percent of the earth's heat. Warmer oceans mean tropical storms and hurricanes that intensify faster, grow bigger, move slower, and dump more water than storms in the past.

Warming oceans acidify the water and deplete oxygen, destroying fragile marine life.

Why should we care?

Global warming disproportionally affects the most vulnerable. People who already live with daily food insecurity are the first ones impacted by rising food prices. People who already live with housing insecurity lose their homes when sea levels rise or when intensified flooding devastates the area. When a water source runs dry, those who depend on it become refugees.

If your lungs are already damaged from breathing polluted air, you are at a much higher risk when a respiratory virus goes pandemic.

The military calls climate change a threat multiplier. Poverty, hunger, disease, access to clean water, political unrest, refugee crises—all are exacerbated by global warming.

Snowball Hoax

In February 2015, the chair of the Senate Environment and Public Works Committee James Inhofe (R-OK) walked to the microphone and pulled out a baggie. From the baggie, Inhofe retrieved a large snowball he had packed himself from the snow outside the chamber.

"Catch this!" said Inhofe, as he threw the snowball into the assembled senators, to no one in particular.

Senator Inhofe, author of the book *The Greatest Hoax: How the Global Warming Conspiracy Threatens Your Future*, smiled as he explained his projectile object lesson. A snowball in February is apparently proof positive that global warming is a fraud.

Despite what all those egghead scientists say, despite the fact that 2015 was the hottest year ever recorded, despite the whole weather-is-not-climate thing, a snowball thrown in the Senate chamber is undeniable proof that global warming is a hoax.

Inhofe is by no means the only politician to call global warming a hoax. It is a common position of political pundits, the guy next door, or the guy at your church.

If it's all propaganda and a hoax, why does global warming get so much attention?

It's about money! Take a look at Al Gore and his famous hockey stick.[8] Al Gore is wealthy. He's obviously profiteering from pushing climate change.

It's much ado about nothing! It's simply the natural cycle of things. It's volcanic eruptions. It's the sun cycles. It's ocean currents.

The science is fake! For every scientist that says human-caused global warming is real, another one says it's not.

It's a flip-flop! Scientists originally told us an ice age was coming, now they tell us the earth is warming—make up your mind.

Texas recently strengthened the state standards for teaching climate science. Previously, Texas earned an F from the National Center for Science Education for its standards addressing climate change. In 2009, Texas had a chance to update the standards, but a state board of education member dismissed climate change as "hooey."[9]

Add "It's hooey!" to the list.

To climate scientists like Jessica Moerman, a fictional global warming would be cause for celebration: "My colleagues and I, if that was true, we would be the first ones to be relieved if this was just some global conspiracy."[10]

Climate and the Evangelical

For many evangelicals, human-caused climate warming is not only hooey; it's downright unbiblical. Robert Jeffress, pastor of the massive First Baptist Church in Dallas and spiritual advisor to Donald Trump, mocked teenager Greta Thunberg following her address at the 2019 United

Nations Climate Action Summit: "Somebody needs to read poor Greta Genesis Chapter 9 and tell her next time she worries about global warming just look at a rainbow," Jeffress said. "That's God's promise that the polar ice caps aren't going to melt and flood the world again."[11]

Katharine Hayhoe is a professor at Texas Tech University and is a world leader in climate science. Her résumé is extensive—she's been named one of *Time*'s one hundred most influential people, one of *Fortune*'s fifty greatest world leaders, and is listed among *Foreign Policy*'s one hundred Global Thinkers. She is the Chief Scientist of the Nature Conservancy.

Katharine Hayhoe is also a committed evangelical Christian. Her husband is a pastor. When she and her husband arrived at their new church in Texas, she met a couple who were thrilled to discover that she studied global warming.

Wonderful! her new friends said. We need someone like you to talk to our children! You would not believe the *lies* they are being taught at school—the ice is melting, polar bears are dying . . .

I imagine Hayhoe with a thought-bubble facepalm.

Two-thirds of white evangelicals reject human involvement in global warming.[12] Given the biblical command to be stewards of creation, this is a puzzle. How did human responsibility in global warming detach from creation care?

Caring for creation in real time is not a problem. Nobody likes polluted lakes, trashy highways, or smog-choked air. Superficially, a polluted environment is inconvenient and unpleasant, right now, to us. We are OK with cleaning up our house.

Evangelicals balk at long-term creation care that demands acceptance of science—particularly science that butts heads with our theological and theo-political beliefs.

Climate versus Genesis

Just as an ancient earth is essential in explaining evolution, an ancient earth is essential in explaining climate change. If your theology demands a literal reading of the first eleven chapters of Genesis, you might have a problem with both areas of science.

Evangelicals have been told that secular scientists' interpretations of climate data cannot be trusted because they are made with the assumption of a billions-of-years-old earth. If the earth is only six-to-ten thousand years old, how can scientists possibly know what the climate was like fifty thousand years ago?

Without a biblical filter, secular scientists inevitably get climate science all wrong: "Hundreds of thousands of years to build up a two-mile-thick ice sheet? No way! Ice sheets were formed quickly in an ice age that followed Noah's flood."[13]

But independent scientific studies agree about the age of the ice! Of course they do. What would you expect from secular scientists, all working from the paradigm of an ancient earth?

Pay no attention to modeling done by climate scientists! Secular models overpredict because they do not take all factors (including a young earth and a global flood) into account. Dramatic climate changes are due to Noah's flood, not human activity. "What you believe about the past determines how you interpret the evidence and draw conclusions"[14] is the young-earth mantra.

And to erase any doubts regarding the slippery slope of old-earth beliefs, Jake Hebert calls climate change "the bitter harvest of evolutionary thinking."[15]

Belief in a young earth demands rejection of climate science.

Faith over Fear

"As long as the earth endures, seedtime and harvest, cold and heat, summer and winter, day and night will never cease." This post-flood promise in Genesis 8:22 is the evangelical proof text for denial of human-caused climate change. Seasons, steadfast and unshakable, are sustained by God's sovereign hand, end of discussion.

For many evangelicals, suggesting that humans can change the climate is an attack on the sovereignty of God. It is the epitome of human arrogance. As they say in Plainview, "There's only one man who knows how much rain we're gonna get and that's God."[16]

If God's sovereignty overrides any damage done to the planet by humans, scientists sounding the alarm look like fearmongers who scare little kids with melting ice and sad polar bears.

Akos Balogh, writing for the Gospel Coalition, sees climate alarmism as fear-driven. Prominent voices on the left like Congresswoman Alexandria Ocasio-Cortez hit the "panic button with an atomic elbow," says Balogh.[17] Climate alarmists are simply the latest in a long line of doomsday prophets calling Christians to live in fear of a nuclear holocaust or an alien invasion.

Like the evangelicals who resisted COVID vaccines, masking, and gathering precautions, climate science deniers refuse to be "controlled" by fear. God is sovereign, so we throw caution to the wind.

Faith over fear in a time of pandemic, faith over fear as the planet heats up.

The Earth Is a Styrofoam Cup

For some evangelicals, addressing global warming is as futile as mopping the deck of the *Titanic*. We've all read the end of the book. We know what happens. It's all gonna burn.

And now (cheery voice), let's all stand and sing: "This world is not my home, I'm just a-passing through."[18]

For many evangelicals, the future of the planet is irrelevant. End-times theology sets our sights on the next world, not the present. We simply pass through this life and let the chips fall where they may. Just ask John MacArthur: "God intended us to use this planet to fill this planet for the benefit of man. Never was intended to be a permanent planet. It is a disposable planet. Christians ought to know that."[19] So use it, abuse it, and throw it away like a Styrofoam cup.

God cursed the earth and is going to fix it all in the end, MacArthur continues, so "we need not repent of the way we have polluted, distorted and destroyed the Creator's work." Let that sink in.

Jerry Falwell, architect of the religious right's flagship organization, the Moral Majority, saw Satan at work in climate science.[20] Not long before his death, Falwell issued a warning: Satan is using environmental activism to distract evangelicals from their true calling—evangelism. It's all part of Satan's plan to put the focus on the creation, not the Creator.

The Dallas Theological Seminary is an influential source of an end-times theology popular in many evangelical circles known as *dispensationalism*. An important aspect of dispensationalism is end-times prophecies, particularly regarding certain geographical regions of the earth.

The late John Walvoord, a renowned dispensationalist at the seminary, approached climate science denial from a different angle: God wants us to have all those fossil fuels. In fact, God *wants* us to burn them.[21] According to Walvoord, it's all in preparation for the end-times. God divinely created massive oil fields in the Middle East, just where they need to be for the coming battles of Armageddon.

Who are we to seek alternative fuels?

Don't Be Bob

Bob Inglis (R-SC) is a one-man cautionary tale. In the super-secret orientation meetings (we can only suppose) that occur between newly elected members of Congress and the party powers-that-be, at least one bullet point says "Don't be Bob."

Inglis served six successful terms in the House as the representative from South Carolina. He initially served three terms, then he kept a campaign promise to bow out after three. He returned to the House six years later and served three more terms.

Inglis was the conservative's conservative from the reddest district in the reddest part of the country. Across the board, he was on the conservative side of every issue, without fail. Inglis easily won his elections with little opposition—until he didn't.

Prior to the race for his seventh term, Inglis let it be known that he believes human-caused climate change is real. Friends ran and allies turned enemy. Not only was Inglis defeated, he didn't even make it to the general election. He lost in the primary by a landslide.

Inglis, an evangelical Christian, understands the religious element to his defeat: "In this district, we call ourselves the 'shiny buckle of the Bible Belt.' So I think for some it is a religious heresy . . . for us to presume that any action that we would take would affect the longevity of his creation. . . . It is an affront to sovereignty of God."[22]

Politically Evangelical

Although two-thirds of white evangelicals reject human involvement in global warming, they aren't alone. White Catholics are right there with white evangelicals in rejecting human-caused climate change.

Which Christian groups are on the other end of the spectrum? Which Christian groups are most concerned with climate change? Near the top are Hispanic Catholics and Black Protestants.

Katharine Hayhoe wants to know why. If all these groups are on Team Jesus, why the divergence? The one determining factor, says Hayhoe, is where a person falls on the political spectrum.[23] Rejection of climate science is most strongly associated with being politically conservative.

It wasn't always so. In the 1970s, conservatives were more likely than self-identified liberals or moderates to express confidence in science. The advent of data on climate change, endangered species, and energy flipped the tables. Scientific data is now suspect to conservatives because it bolsters arguments for more government regulation.[24]

Paul Douglas is a popular speaker, author, and long-time broadcast meteorologist in the Twin Cities area. Douglas is a Christian and a self-described fiscally conservative, small-government Republican. Douglas is doing his best to convince fellow Christian conservatives that human-caused climate change is real. It used to be, says Douglas, that conservatives were in favor of conserving. Apparently, conserving-conservatism doesn't apply to the planet.[25]

The power of the politically evangelical is not lost on politicians.

In his bid for the presidency, former Texas Governor Rick Perry featured climate-change denial on the campaign trail. It's not the fault of humans, said Perry. The science isn't proven, he said.[26] Why should he, a good conservative, put our children's futures at risk on the say-so of some scientists?

And remember the libertarian think tank that drafted the Great Barrington Declaration during the COVID pandemic? Applauded by evangelicals, the Great Barrington Declaration was a petition calling for a halt to pandemic shutdowns for economic reasons. The pandemic petition, however, was not the think tank's first foray into conservative politics and science. They have a history of rejecting climate science and denying environmental health risks, also for economic reasons.

Many evangelicals are in favor of limited government, but not for the expected libertarian government-off-my-back reasons. There is a belief that limited government is God-ordained and that authority should reside primarily in the church, the family, and in individuals.[27] Policies addressing environmental issues are seen not only as government overreach but also as an attack in an ongoing culture war.

Using the strong arm of the Moral Majority, Jerry Falwell encouraged evangelicals to oppose environmental protection legislation because regulations suppressed the free market. Capitalism, according to Falwell, is the biblically endorsed economic system. As such, free markets and businesses unencumbered by government regulation are our best defenses against communism.[28]

To John MacArthur, climate science is both political and a front in the culture war, with some misogyny thrown in for good measure: "A hundred billion dollars has been spent to make a case for global warming. . . . You say, well what motivates this. . . . I think it's driven by the socialist mentality that resents success, even some of the feminist mentality that resents male success."[29]

Don't Make Me Solve the Problem

Denis Hayes was raised in the paper mill town of Camas, Washington. It was the 1950s, and the mill smokestacks billowed out unfiltered sulfur dioxide and hydrogen sulfide into the air. The whole town smelled like rotten eggs, but to the residents of Camas, it was the smell of prosperity and progress.[30]

When it rained (which it did almost daily in southwest Washington), the sulfur dioxide mixed with the rain and formed an acid. Acid rain pitted the roofs of cars, destroyed the roofs of houses, and gave everyone a sore throat.

But to question the mill was to question your paycheck.

Instead of cleaning up the air, the paper mill simply installed a car wash at the exit of the parking lot. Problem solved.

Katharine Hayhoe is passionate about climate solutions. When conversations start out with "God's in control" or "The earth is temporary," Hayhoe says they quickly divert to "I don't want another tax," "Regulations will destroy the economy," and, of course, "I won't be able to drive my truck anymore!"[31]

The fact is, says Hayhoe, we fear the solutions more than we fear the impact of global warming. We fear solutions will limit our personal freedoms or lower the quality of our lives. We fear more government regulation. Impacts of global warming, on the other hand, are either far away in the future or far away in another country. My truck is in my driveway, right now.

Solution aversion makes it easier to dismiss or deny scientific evidence. If we deny the science, we don't feel like a bad person for failing to act.

If we accept the science, we feel obligated to do something about it.

It's a quandary that boxes evangelicals into a denialist corner.

Same Song, Third Verse

When climate change entered our evangelical consciousness, who were the messengers? It was the same pointy-headed scientists who have been on the other side of the fence regarding evolution, the age of the earth, and a global flood. And when COVID entered the picture, infectious disease scientists were standing right there with them.

We know what scientists do. Scientists hide evidence. Scientists massage the data. Skeptics are silenced. Culture war, fear, rights, freedoms—we've been through it all before.

In climate science denial, we sing the third verse of a familiar song. The climate science verse sounds very much like the evolution verse and the pandemic verse, with repeating patterns of denial, just different areas of science.

Evangelicals who attack Katharine Hayhoe's climate science posts on her social media are usually antimasking on theirs. I see the same trend on my own Facebook feed. A minister who regularly posts anti-evolution and anti-COVID memes wrote this on his wall: "These are the same folks pushing human-caused global warming . . . Talk about faith! I'll stick with Scripture and a Creator, thank you! It takes way less faith! Follow the science they say! What science! This is fantasy." Antagonism toward fact-based decision-making is often part of a package deal.

The writers at Answers in Genesis demand absolute proof for evolution, and climate science is held to the same standard. When climate science is really settled, they say, the evidence will be overwhelming. Even though evidence for human-caused global warming *is* overwhelming, Answers in Genesis says it's not.

When your standard is "absolute proof," you'll never get there.

When the demand is for absolute proof, there's never enough evidence, and the evidence we have is sketchy. So sketchy in fact, we flood our state houses with legislation to "teach the controversy" of climate science, just like we did with evolution.

Don Kopp, a representative in the South Dakota House, is an evangelical Christian and quite vocal about his opposition to evolution.[32] In 2010, Kopp successfully led the state to adopt a resolution calling for "balanced and objective" presentations of climate science in public schools. The language in the South Dakota climate science resolution was almost identical to the language used in the 1980s Supreme Court case in which teaching creationism was ruled unconstitutional.

It was not a coincidence.

The Discovery Institute, a Seattle-based intelligent design think tank, has long supported a "strengths and weaknesses" approach to teaching evolution. Discovery also provides models for drafting "balanced approach" climate science legislation at the state level and works for passage of the bills.[33]

In state houses across the country, similar "balanced approach" climate science bills and resolutions were entered. Some succeeded, some failed. The wording varied, but the intent was the same. Teach the controversy. Teach both sides. Teach the strengths and weaknesses.[34]

Such bills and resolutions imply that climate science and evolution are just "ideas," that "both sides" have facts, that all sides are equal.

Arizona State Representative Judy Burges sponsored a "balanced approach" climate science bill in her state: "There should be an opportunity for teachers to step up to the plate and give their opinion . . . without retribution."[35]

Representative Burges, I hear Flatrick Burke is looking for a geography teaching job.

Balanced treatment bills are a Trojan horse, says Mark McCaffrey of the National Center for Science Education. Balanced treatment bills allow teachers to present "all sides" and act as if there is a scientific controversy, when in fact, there is none.[36]

And finally, taking a page from the COVID science denial playbook, is Jake Hebert of the Institute for Creation Research (ICR). Implanting a microchip via a vaccine was just the beginning. In an anti-climate science podcast, Hebert enumerates solutions being proposed by scientists to combat global warming.[37]

"They" want to bioengineer people to make them shorter, so we won't need as much food. They want to bioengineer people to take away the desire for meat. They want to bioengineer us to have eyes like cats, reducing our need for light and our carbon footprint at the same time.

Beware, says Hebert. Climate scientists want to turn your children into miniaturized vegetarian cat people.

11

Hello, Dolly!

She didn't have a shirt pocket, so embryologist Karen Walker tucked the little container holding the tiny egg inside her bra to keep it warm on the chilly trip from the farm to the lab.

It's not unusual for it to be chilly in Scotland, but it is a bit odd to find a cell laboratory tucked in a corner of a Scottish sheep farm. "Laboratory" is a bit of a stretch: it was actually a cupboard, just big enough for two chairs and an incubator.

On a cold February morning in 1996, Dolly the sheep was cloned from an adult sheep living on a farm near Edinburgh, Scotland.

Dolly wasn't just the first mammal ever cloned. Dolly was the first *animal* ever cloned from an adult cell of an existing animal.

An egg from one of the farm's Scottish Blackface sheep (the one carefully incubated by Walker) was brought into the tiny lab. In a delicate process, Walker and her partner, Bill Ritchie, removed the nucleus from the egg. Egg nuclei, like the nuclei of all cells, contain the DNA of the organism.

Likewise, the researchers removed the nucleus from a mammary cell taken from the udder of a second adult sheep, a white-faced Finn Dorset. The mammary cell nucleus (containing DNA) was then inserted into the empty egg.

Next, the researchers implanted the egg with the new DNA into a surrogate, another Scottish Blackface sheep.

Nobody really expected success. Nobody was terribly confident that the DNA from an adult mammary cell could "reprogram" an egg to fashion the wide variety of cells found in an entirely new animal.

But implantation was successful. The surrogate was pregnant. And at the end of a normal pregnancy, the surrogate gave birth to a healthy lamb.

Walker was away at a wedding at the time, so Ritchie sent a birth-announcement fax to her hotel: "She has a white face and furry legs!"

I bet the hotel staff thought: "Well. . . . That's a, umm, *unique* baby. . . ."

Only the DNA donor was a white-faced sheep. Both the surrogate and the (empty) egg donor were black-faced. Genetic tests would later confirm what appearances first revealed: Dolly was a clone of the sheep who donated the mammary cell DNA.

Dolly was a healthy ewe who went on to birth a total of six lambs. Dolly was euthanized at age six due to a lung disease she and two others in the flock had developed.

Because of her DNA origin in adult mammary cells, researchers named Dolly (the sheep) for Dolly Parton (the singer). No disrespect for Ms. Parton was intended, according to the researchers, and the singer's agent supposedly responded: "There is no such thing as baaaaaaaaad publicity."[1]

Dolly is on display behind glass in the National Museum of Scotland—behind glass because people kept nicking bits of her wool. I have a fun photo of me with the (second-most) famous Dolly!

Lessons from a Little Lamb

If we've learned any lesson from evolution science, climate science, and pandemic science, it's this: science marches on. We can deny it, we can ignore it, we can misrepresent it, but science isn't going anywhere. Scientists keep collecting evidence, tweaking theories, and adding to our understanding of the physical world.

The twenty-first century is emerging as a century of biotechnology, ready or not.

Following the announcement of Dolly's birth, reactions ranged from the hopeful ("New cures for diseases!") to the fearful ("Oh no! Armies of cloned humans!"). In reality, Dolly's birth did not have much impact on animal cloning. Aside from the prize racehorse or prize bull here and there, cloning did not become a big deal.

If we've learned any lesson from evolution science, climate science, and pandemic science, it's this: science marches on. We can deny it, we can ignore it, we can misrepresent it, but science isn't going anywhere.

Before Dolly, we thought that adult cells, once they had matured and developed into their final form (like heart cells, liver cells, nerve cells, etc.), were stuck in their final form and could not regress back to their unspecialized embryonic state.

Dolly showed us that a specialized adult cell can be reprogrammed into an unspecialized embryonic cell. Unspecialized cells, capable of becoming any of the specialized cells in an animal's body, are called *stem cells*, and stem cells are gold in medical research.

Dolly was a landmark lamb. Dolly inspired scientists to pursue human applications in biotechnology using cloning techniques. Thanks to Dolly, cell biologist Shinya Yamanaka began developing stem cells from adult mice cells, a feat that won him a Nobel Prize in 2012: "Dolly the Sheep told me that nuclear reprogramming is possible even in mammalian cells and encouraged me to start my own project."[2]

Stem Cells

Stem cells are essential in drug and disease research. There are two types of stem cells: embryonic stem cells and adult stem cells.

Embryonic stem cells are formed when, after a mammalian egg is fertilized, the egg begins to divide. After about four to five days, it is a hollow ball of cells, smaller than the dot over this *i*. The outer layer of cells will eventually form the placenta, and the inner cells will eventually form the embryo.

At this point, the inner cells are unspecialized. Ultimately, the inner cells will specialize into all the cell types and tissues of the body—muscles, blood, nerve, bone, connective. But now, just days after fertilization, they are unrestricted free agents—undifferentiated, generic *stem* cells, capable of becoming any type of cell in the body.

Embryonic stem cells are obtained from unused embryos left over from in vitro fertilization procedures. All embryos donated for research must have full informed consent from the donors. All unused embryos, if not donated for research, are destroyed.

Embryonic stem cells cultured in a lab will grow into colonies of hundreds of thousands of cells. In fact, they can continue dividing for years and years. Embryonic stem cell lines used in research are not retrieved directly from an embryo but are copies of the original cells.

There are two types of adult stem cells. One type comes from fully developed body tissues like brain, skin, and bone marrow. In fully developed, specialized tissues, there will be a small number of stem cells for *that type* of tissue. For example, a stem cell from the liver will only make more liver cells. Stem cells associated with a specific tissue are used to replace cells lost through injury, disease, or general wear and tear.

Blood stem cells are found in the bone marrow. Blood stem cells can transform into all the different types of blood cells. When blood cells are damaged or destroyed by cancer treatments, for example, bone marrow can be transplanted to replenish the blood cell population.

The second type of adult stem cell is an *induced* stem cell. Induced stem cells are adult cells changed in the lab to act like embryonic stem cells. Dolly taught us this is possible!

DNA taken from an adult body cell is placed into an unfertilized egg which has been stripped of its nucleus. Doing this reprograms the egg cell into an embryonic state. When the cloned embryo reaches the cell-ball stage, the inner cells can be removed and grown into an "induced" stem cell line.[3]

Although induced stem cells are very much like embryonic stem cells, we haven't found one yet that can produce every kind of cell and tissue.

Stem Cells in Disease

People who suffer from severe arthritis often depend on painkilling drugs to manage their disease. An unfortunate side effect from many of these drugs is gastrointestinal problems like ulcers and gastrointestinal bleeding.

In 1999, the painkiller Vioxx was launched and promised to be a lifesaver for arthritis sufferers. Clinical trials showed that Vioxx was much safer on the digestive system than older drugs.

Unfortunately, a significant number of people taking Vioxx suffered heart attacks, and many died. For most people, Vioxx was a safe alternative. But for others, there were serious cardiac side effects.[4]

Even the largest clinical trials cannot include every possible human genetic combination. What if there were a way to try a new drug on a huge number of people with diverse genetic makeups? What if there were a way to test a new drug on human cells before the drug entered clinical trials?

Drugs typically go through intensive animal trials for years before they are given to people. Even if a drug appears to be perfectly safe in animals, there is no guarantee the same will be true for humans. Rats and mice are commonly used in medical research, but they do not perfectly mimic human reactions to drugs.

Already we can test some drugs for heart toxicity in induced stem cells.[5] Testing new drugs on stem cells before testing in human trials means substantially less risk to patients.

Using data from the Human Genome Project, Susan Solomon's foundation is developing a technology that will allow the creation of stem cell "avatars." The technology has the capacity to produce thousands of cell lines, representing a wide variety of human genetics. Screening drugs using technology of this sort could help us avoid another Vioxx-type tragedy.[6]

Stem cells also hold promise as replacement tissues in treating spinal cord injuries and eye diseases. Therapies using embryonic stem cells are currently in human clinical trials.

Induced stem cells from adult cells are also being tested for therapies, but induced cells do not reprogram as well as do embryonic stem cells. However, induced stem cells have the advantage of being genetically matched to the patient, reducing the risk of tissue rejection.[7]

Fetal Tissues

Fetal tissue studies have produced landmark research in medicines, vaccines, and in the understanding of a multitude of diseases.

Fetal tissue for research is obtained from both spontaneous and elective abortions. Informed and written consent is required for each donation, and in all cases, the fetal tissue would be destroyed if not donated.

Cells are isolated from tiny pieces of fetal tissue. Each cell can divide up to fifty times. Then each of those cells divides, and so on. From just a few original fetal cells, tens of millions of cells are produced.

Cell biologist Leonard Hayflick has ten million frozen fetal lung cells, all derived from just a few original cells. He has enough cells to provide the world's vaccine manufacturers with cells for several years.[8]

Hayflick also created WI-38, the oldest fetal cell line in use, from an elective abortion in Sweden in the early 1960s. Two other widely used lines are a kidney cell line from 1972 and a retinal cell line from 1985.

Hayflick's lung cell line is currently used to produce vaccines for varicella, rubella, hepatitis A, and rabies. In fact, fetal cell lines are a key component in producing most conventional vaccines.

Fetal Cell Lines and Vaccines

Conventional vaccines deliver small doses of a weakened or inactivated virus, giving your body a preview of the virus without making you sick. Your immune system builds a memory of the virus, so if an actual full-strength virus attacks, your body will be ready to destroy it.

Here's the challenge—viruses can only grow and reproduce inside a host cell. Vaccine manufacturers need a way to grow mass quantities of viruses for inactivation. Fertilized chicken eggs are one answer and are commonly used to grow viruses for flu vaccines.

Viruses grown in fertilized chicken eggs, however, can mutate. Scientists prefer to grow viruses in mammalian cells, but this is still problematic. Animal cells can harbor undesirable viruses that can contaminate the vaccine. An early version of the polio vaccine was found to be contaminated with a monkey virus.

The risk of contamination is greatly reduced by growing viruses in human cells—specifically, human fetal cells. Adult cells, with a lifetime of exposure to viruses, harbor potentially contaminating viruses.

Fetal cells lines were used in the production and testing of COVID vaccines.

The Johnson & Johnson vaccine is a viral vector vaccine.[9] Vector vaccines use a nonreplicating, harmless virus to deliver genetic instructions to a human cell. The Johnson & Johnson vaccine uses a harmless adenovirus, not the COVID virus, to deliver the instructions for building a COVID spike protein. A retinal cell line isolated in 1985 is used to grow the vector viruses.[10] The Johnson & Johnson vaccine contains no actual fetal cells—all cells are extracted and filtered out.

The Pfizer and Moderna vaccines use messenger RNA, wrapped in an oily microbubble, to deliver the instructions for building a COVID spike protein. Early in the development of mRNA vaccines, a fetal kidney cell line from 1973 was used to see if the technology worked.[11] We needed to know if human cells would actually take up the mRNA and build a protein. No fetal cells or tissues are used in the actual production of mRNA vaccines.

Drug and Disease Research

Fetal cell lines are also used in the development of common drugs like acetaminophen, ibuprofen, and aspirin. In addition to over-the-counter drugs, fetal cell lines are used to make drugs for hemophilia, rheumatoid arthritis, and cystic fibrosis.[12] An experimental antibody treatment used to treat COVID was developed using fetal kidney tissue.[13]

In addition to vaccine and drug development, fetal cell lines are critical in disease research. Much of what we know about the causes, prevention, and treatment of Parkinson's disease, Huntington's disease, Zika, human immunodeficiency virus (HIV), diabetes, blindness in premature infants, macular degeneration, diabetes, cancers, birth defects, and heart disease comes from fetal tissue research.[14]

Regulation Ping-Pong

Federal regulations regarding embryonic stem cells and fetal tissues have oscillated between restrictive and less restrictive for more than two decades, depending on the political party in charge.

In 2001, George W. Bush banned federal funding for creation of new human embryonic cell lines.[15] All embryonic cell lines created before that date would be eligible for funding, but only twenty-one cell lines proved to be useful for NIH investigators. Unfortunately, the approved cell lines were not genetically or ethnically diverse, limiting their usefulness.

Less than two months after he took office, Barak Obama revoked the Bush administration's embryonic stem cell restrictions. Researchers could use embryonic stem cell lines previously restricted under the Bush administration, but federal money could not be used to create new cell lines.

Donald Trump threatened embryonic stem cells but acted on fetal tissue research.[16] His administration banned fetal tissue research by in-house NIH scientists but allowed funding for nongovernment scientists. Even so, grant applications for new fetal tissue research added layers of supplemental requirements and lengthy reviews, effectively adding months to the process.

Ironically, the experimental antibody treatment given to President Trump when he was hospitalized with COVID was developed using a fetal tissue cell line.

Months after he took office, Joe Biden lifted most of the Trump administration's fetal tissue research restrictions.

Obviously, research using embryonic stem cells and fetal tissues must be regulated, but politics made science into a game of ping-pong.

A Moral Dilemma

Even as infections soared, evangelicals reached into their grab bag of COVID science denial and pulled out a seldom-used dodge: religious exemption.

Before the advent of COVID-19, evangelical opposition to stem cell and fetal cell research was generally hypothetical. Politically conservative, antiabortion evangelicals opposed the research due to the source of the cells. Conservative politicians, courting the support of evangelicals, were more than happy to oblige.

Social conservatives have long opposed funding research with embryonic stem cells and fetal tissues with taxpayer money.

At issue is whether it is morally acceptable to use a human embryo, destined for destruction, in medical research. At issue is whether it is morally acceptable to use a lifesaving vaccine, and for that matter, a myriad of other drugs and treatments dependent on research using tissues harvested from decades-old abortions.

Yamanaka's Nobel Prize-winning feat creating induced stem cells from adult cells has reduced the need for ethically problematic embryonic stem cells and fetal tissues.[17]

The need is reduced, but not eliminated: "Fetal tissue has unique and valuable properties that often cannot be replaced by other cell types," says the International Society for Stem Cell Research. It "remains the gold standard for evaluating the accuracy of models of human fetal development."[18]

Sally Temple, scientific director of the Neural Stem Cell Institute, calls fetal tissues "essential" and says that to suggest otherwise is "simply wrong."[19]

Notably, the Roman Catholic Church, historically opposed to abortion and leery of the association with fetal cell lines, officially endorsed the COVID vaccines. The Pope, as well as Catholic cardinals and bishops, supported vaccination as a "moral obligation."[20]

Where Are the Evangelical Voices?

Curtis Chang is an evangelical theologian, seminary professor, and former pastor. Chang created a website and a video series called *Christians and the Vaccine.* Speaking as an evangelical to evangelicals, Chang attempted to allay fears about COVID vaccines and fetal tissues. Chang understood the hesitancy—even if the vaccines did not contain fetal cells, fetal tissues were present in development of the vaccines. Some evangelicals did not want the guilt by association.

Chang lives on the West Coast, where an important means of transport for large goods is rail lines. As a Chinese American, Chang is aware of the history of the rail lines entering and leaving California, rail lines following the pathway of the original transcontinental railroad.

As a Chinese American, Chang is also acutely aware of the racism, lynchings, and massacres associated with building the transcontinental railroad: "But that doesn't make those current railroad lines guilty, nor the fact, or me guilty by receiving those things. It's just an inevitable result of the fact that we live in a fallen world and that we're all ensnared in touching lines that if you'd go back far enough, you can trace to some wrongdoing."[21]

"Today's cell lines," says Chang, "are like those railroad tracks. They've been laid down on tracks, that if you go back far enough, you can trace to some wrongdoing." But today's cell lines, concludes Chang, are not guilty.[22]

Chang is a theologian, but he is not a scientist. Where are the evangelical voices speaking to the science? Where are the evangelical voices in science speaking to the use of decades-old cell lines? Where are the evangelical voices speaking to creation of new cell lines? Where are the evangelical voices in science weighing the reduction of human suffering and use of tissues already marked for destruction?

Where are the evangelicals talking about redeeming the circumstances?

Well, Francis Collins tries. Francis Collins's scientific credentials are impeccable, even to his colleagues who cast a side-eye at him because of his evangelical faith when he was nominated to head the NIH.

There are things we simply would not know if it weren't for research done with fetal tissues, says Collins, the scientist. And for Collins, the Christian scientist, it is also a matter of ethics: "Even for somebody who is very supportive of the pro-life position, you can make a strong case for this being an ethical stance. . . . If something can be done with these tissues that might save somebody's life downstream, perhaps that's a better choice than discarding them."[23]

It's an issue with which he wrestles.[24] As someone "who really does think that human life is sacred," Collins asks us to consider the ethics of using fetal tissue to help someone, rather than throwing it into the incinerator.[25]

Even as he was speaking up for science as an evangelical Christian, Collins was savaged on the popular conservative website *Daily Wire* and other conservative, antiabortion sites for his role in fetal tissue research.[26] According to many conservative media outlets, Francis Collins is actually a bad guy who is hoodwinking Christians, despite his "aw shucks" personality and Mr. Rogers looks. His offense? As head of the NIH, Collins funded research using abortion-derived fetal tissue. Collins supports the research. What's more, Collins thinks the research is important.

He also supports COVID science, and he's for masking and vaccines, but that's not the worst of it. Collins accepts evolution. In fact, he started BioLogos, an organization that encourages acceptance of evolution by people of faith. When you read about evangelical disdain for Collins and his support for NIH fetal cell research or his leadership in the COVID pandemic, it is invariably tied to his acceptance of evolution.[27]

What Else?

In the 1950s, James Watson, Francis Crick, and Rosalind Franklin worked out the physical structure of a DNA molecule and opened the floodgates of modern genetics. Medicine, agriculture, and biotechnology leaped forward, and the twenty-first century took the stage as the century of genetics.

Far beyond simply describing and explaining DNA, we can now map it, alter it, edit it, and use it.

Gene Therapy

For the first time in history, the 2020 Nobel Prize in Chemistry was shared by two women, Jennifer Doudna and Emmanuelle Charpentier.[28] Doudna and Charpentier were intrigued by an ancient immune response discovered in bacteria. When a virus infects certain bacteria, the bacteria react by chopping up the DNA of the invading virus. The bacteria keep pieces of the chopped-up viral DNA and incorporate it into their own DNA, kind of like a genetic "mug shot" of the invader. If the virus attacks again, the bacteria send chemical "scissors" to cut up the viral DNA at the precise location of the mug shot DNA.

This bacterial immune response is named for the viral mug shots: Clustered Regularly Interspaced Short Palindromic Repeats (CRISPR, pronounced like the drawer in the refrigerator where you store your vegetables). Doudna and Charpentier retooled CRISPR and used it to edit DNA in living cells with incredible precision and accuracy.

Already Doudna and Charpentier's discovery has impacted plant breeding and agriculture. Most exciting, however, are the possibilities for treating genetic diseases.

Victoria Gray, a wife and mom of three, was born with sickle cell disease, a devastating genetic disorder. One single mutation in the DNA of people with sickle cell disease turns their red blood cells into deformed, sickle-shaped cells. The misshapen cells get stuck in blood vessels, resulting in damaged organs, debilitating pain, and often premature death.

The gene that causes sickle cell disease produces a defective form of hemoglobin, the protein in red blood cells responsible for carrying oxygen. There is another kind of hemoglobin made by babies before birth called *fetal hemoglobin*.

After a baby is born, a gene called BCL11A turns on. BCL11A tells blood cells to stop making fetal hemoglobin and start making adult hemoglobin. In people like Gray with sickle cell disease, adult hemoglobin is defective.

In 2019, doctors removed cells from Gray's bone marrow. Using CRISPR technology, the BCL11A gene was cut out. Doctors then infused the modified cells back into her body, restoring production of fetal hemoglobin.[29] A year later, almost 50 percent of the hemoglobin in Gray's system was fetal hemoglobin, enough to significantly reduce her pain and hospitalizations. Victoria Gray continues to thrive.

CRISPR therapy holds promise for other genetic diseases, including beta thalassemia, cystic fibrosis, and some cancers.

"It's a blessing," Gray said. "It gave me hope when I was losing it. So I feel joy, you know, knowing that there is hope."[30]

Three-Parent Babies

Mitochondria, nicknamed "the powerhouse of the cell," are the structures in our cells that convert the food we eat to energy. When mitochondria are defective, it is not surprising that high-energy needs systems are impacted: the central nervous system, the respiratory system, the heart, the muscles, and the eyes.

Mitochondria are unlike any of the other tiny structures floating around in the cytoplasm of our cells. Mitochondria have their very own tiny little genomes—their very own collection of thirty-seven genes not found in the nucleus of the cells. This tiny little genome controls the critical energy-generating functions of the mitochondria.

It's generally understood that half of a person's genes come from biological mom and half from biological dad. This is not true. Everyone has slightly more DNA from mom.

Both eggs and sperm have nuclei; both eggs and sperm have liquid cytoplasm. *But* at the point of fertilization, *only* the nucleus of the sperm enters the egg to fertilize it. What's the result? A fertilized egg with nuclear DNA from both egg and sperm, but cytoplasm from only the mother. And because mitochondria are found in the cytoplasm, mitochondria are all from mom.

In 2016, the birth of the world's first three-parent baby was announced.[31] The baby's mother carries a genetic mutation in some of her mitochondria for Leigh syndrome, a devastating and fatal disease. Although the mother was healthy, she suffered multiple miscarriages and lost two children to Leigh syndrome.

An egg was retrieved from the mother. The nucleus was removed from the mother's egg. Next, the nucleus was removed from a donor egg with healthy mitochondria. The nucleus from the mother was then placed in the donor egg, which now contained only cytoplasm. The result? Nuclear DNA from the mother, mitochondrial DNA from the donor.

The egg was then fertilized in vitro with sperm from the father. The result was a healthy baby boy with DNA from his mother and father, and a tiny fraction of mitochondrial DNA from a donor egg.

Recombinant DNA

Beta-carotene, a precursor of vitamin A, is not produced in the edible parts of a rice plant. In areas of the world with heavily rice-dependent diets, vitamin A deficiency is the primary cause of blindness and much disease.

Recombinant DNA technology allows us to insert genes from daffodils and bacteria into ordinary rice. The resulting plant produces rice grains with a golden color—golden because they contain beta-carotene.

Fears and misinformation regarding GMOs kept golden rice out of production for three decades. Finally, in 2021, golden rice was approved for consumption by the Philippines and five other countries, including the United States.[32]

Recombinant DNA technology allows us to insert the gene for human insulin into bacteria. We can then grow vats and vats of these bacteria, all churning out insulin, *human* insulin. Before 1980, people with insulin-dependent diabetes relied on insulin harvested from cows and pigs, which was often inefficient and could potentially trigger an allergic reaction.

In agriculture, recombinant DNA technology allows us to insert a gene for a natural insecticide into crops, greatly reducing the need for external pesticides.

We could go on and on for chapters and chapters.

If we've learned anything from seven *Jurassic Park* movies, it's that new science with cells and DNA and technology can be complicated. Maybe not rampaging-tyrannosaurs-complicated, but a world of cloning, stem cells, and DNA manipulation can take your breath away.

We can deny it; we can ignore it; we can misrepresent it. Better, we can learn about it.

Where are the evangelical voices speaking to the science?

Will we listen?

12

Living as People of Faith in a Modern Scientific World

It was a stretch to say the least, more like a giant leap. Unfortunately, it was a leap into the unknown.

Although the global scientific community applauded the application of CRISPR technology to cure sickle cell disease, Dr. He Jiankui's application of CRISPR was met with shock and condemnation—and prison. A Chinese court sentenced He, a biophysicist, to three years in prison for "illegal medical practice."[1]

In 2018, He used CRISPR technology to modify the DNA of human embryos. He disabled a gene in the embryos that allows HIV to enter cells. The embryos were then implanted into two women, resulting in the births of twins and a single baby.

He flouted regulations and ethical guidelines in both China and worldwide. Research in human embryo gene editing is far too preliminary to leap as He did into the unknown.

Gene editing to treat sickle cell disease, as with Victoria Gray (see chapter 11), is done in body cells, not embryos. Unlike gene edits in embryos, gene edits in body cells cannot be passed to offspring.

He edited a gene called CCR5, a gene involved in HIV infection, in the genomes of healthy embryos. However,

CCR5 is also related to major brain functions. He induced a mutation with little benefit and with potential harm.

"Science can tell you what you can do and how to do it," says evangelical pastor and author Tim Keller, "but it can never tell you whether or not you ought to do it."[2] Clearly, He's experiment with human embryos was ethically problematic. Ethics should govern what we do with science, and scientists should be on the frontline, leading the charge.

Changing Lanes

In the fall of 2018, emergency care doctors, trauma surgeons, and neurosurgeons flooded social media with disturbing photos and wrenching stories: Bloody operating room floors. Emergency rooms littered with sterile wrappings of equipment after a failed attempt at saving a life. A blue plastic chair where the doctor sat when she told parents their child had died. The photos all bore the same caption: "This is my lane."

In response to an article on firearm violence in the *Annals of Internal Medicine*, the National Rifle Association (NRA) tweeted: "Someone should tell self-important anti-gun doctors to stay in their lane."[3]

The NRA tweet set off a storm of posts and emails from doctors. This IS my lane, the doctors said: "COME INTO MY LANE. Tell one mother her child is dead with me, then we can talk."[4]

It's foolish to think that those who contend with the aftermath of gun violence have nothing important to say about guns.

Sometimes lanes need to inform other lanes.

One of the most influential evolution biologists of our time, Stephen Jay Gould, introduced us to the term *non-*

overlapping magisteria. Although a very lofty-sounding term, it was simply Gould's way of saying that science and religion do not conflict.

Science and religion, according to Gould (an agnostic and a secular Jew), exist in separate areas of authority, areas which do not overlap. As long as science stays put in its "magisterium" and religion stays put in its "magisterium," all is well and we play nicely together.

It's an interesting concept, and one that evokes deep philosophical discussions. Do science and faith occupy two wholly separate and distinct worlds, never intersecting, never speaking to each other? Do science and faith need to stay in their respective lanes? Or do faith and science have something of value to say to each other?

Conditioned to Distrust

In a creationist paradigm, science has nothing to say outside of a literal, historical reading of the Bible. Henry Morris, the architect of the modern creationist movement, pulls no punches: "When science and the Bible differ, science has obviously misinterpreted its data."[5] In other words, science has nothing to add to the conversation, at least nothing in the area of origins. We start with a young earth or an instantaneous creation, and we retrofit scientific evidence to fit it. We are conditioned by creationism to approach science with foregone conclusions.

Almost thirty years ago, Mark Noll succinctly defined the problems generated by creationism in his monumental work, *The Scandal of the Evangelical Mind*. "Creation science has damaged evangelicalism by making it more difficult to think clearly about human origins, the age of the earth, and mechanisms of geological or biological change," observed

Noll.[6] What's more, says Noll, creationism has profoundly damaged the ability of evangelicals to look at the natural world, in general, and to interpret what we see.

Fighting evolution for so long shaped the way evangelicals approach science. We are conditioned to distrust. When we don't understand the details, we listen to familiar voices telling us it's not true. We listen to those who think like us—theologically and theo-politically.

We found it easier to blame evolution for atheism and a plethora of other cultural ills than to falsify it. When science is a battleground for a culture war, scientists are the enemy. Scientists aren't like us. They don't share our values. We turn, then, to our group, to those who *do* share our values.

John James Kirkwood is cohost of the *Voices in the Wilderness* podcast and is a self-described former culture warrior pastor. "I had my authorities," says Kirkwood. "I felt I was well read on the subject, but I was really only reading within confirmation bias."[7]

Having foregone conclusions, we define truth not by expertise but by what's accepted by our people.

Speaking to Facts with Faith

We can fear science; we can deny it; we can vilify it. Or we can lead.

We can choose to look away, but the earth is not flat, the climate is warming, evolution is real, and the scientific method works. Given the facts, how will our faith inform our response? In a modern scientific world, who is going to speak as an evangelical to evangelicals? And who is going to speak with an evangelical voice to the wider world?

Science can give you the facts, but science does not provide moral guidance as to how we should respond to the facts. Instead of insisting on our own set of alternative facts, we could be leading with a faith-informed voice.

Ben Stein is an American author, actor (Bueller? Bueller?), and political commentator. Stein cowrote and starred in the 2008 documentary-style film *Expelled: No Intelligence Allowed*. It grossed almost eight million dollars—a tidy sum for a documentary.

In the film, Stein argues against evolution as the explanation for the diversity of life on earth. Stein claims a conspiracy within the scientific community to exclude any consideration of intelligent design creationism. What's more, according to Stein, evolution is morally bankrupt and leads to things like eugenics and the Holocaust.[8]

Instead of insisting on our own set of alternative facts, we could be leading with a faith-informed voice.

Evangelicals loved the film. Stein received the Phillip E. Johnson Award for Liberty and Truth from Biola University, an evangelical university in Southern California. In his acceptance speech, Stein said this: "Under the Darwinist paradigm, life is meaningless. Under the Darwinist paradigm, we are just mud."[9]

Instead of "Evolution means life is valueless" and "Evolution means we're just animals so we can do as we please" and "Evolution means there's no need for God," what if we spoke words of faith in the context of scientific facts?

Here's an alternative scenario:

We have the facts. Humans evolved. Humans share ancestry with all life. Accepted. Now, let's talk about the meaning of a life in Christ. Let's talk about overcoming our "me-first" survival instincts in order to love our neighbor. Let's talk about what it means to be chosen as God's image bearers in creation.

We have the facts. Humans are causing the climate to warm, with devastating consequences for the most vulnerable among us. How can we incorporate long-term creation care into our theology of stewardship, right now, in our time?

We have the facts. Germ theory is real. In a public health crisis like a pandemic, do we speak words of truth and evidence and facts, or do we speak words of conspiracy and misinformation?

In a public health crisis, how do we best spend our resources? How do we spend our personal resources? Our church resources?

Do we spend our money paying legal fees so we can keep our church buildings open in a pandemic? Or do we support communities where few people can work from home? Where many families live in crowded homes with little room to socially distance? Where business shutdowns mean loss of income and unpaid bills? Where is the evangelical voice in resource allotment during a public health crisis?

Biotechnology and modern genetics aren't going anywhere. How are evangelicals speaking to these cutting-edge areas of science?

Should we use gene editing only to treat a sick person? Or should we use it to prevent children from being born with devastating conditions like Tay-Sachs, cystic fibrosis, or Huntington's disease? Should we only correct a disease, or is it OK to enhance health in other ways?

Given the facts, how might faith inform our decisions?

The Damage of Denial

Jessica Moerman felt like she didn't belong.

As a young Christian, Moerman felt called to do good in the world. She felt called to partner with God in doing so.

Growing up evangelical, Moerman saw faith and science as enemies. But what do you know? She fell for earth sciences and became a climate scientist, fully intending to honor Christ with her work.

Looking around the scientific community, Moerman realized that there weren't many like her—both scientist and evangelical. Maybe evangelicals don't belong in science, she thought. Maybe the broader scientific community wouldn't accept her.

Still, she persisted. Moerman wants to be an example to young Christians who, like her, hesitate to choose science as a calling that is God-honoring. Here's Moerman: "We don't have those messengers in science who can then speak to the Christian community with authority. We don't have those trusted messengers who can speak to conservatives with authority because for whatever reason, there's a feeling that you're not welcome in the sciences."[10]

Vilifying science deprives the field of bright evangelical minds. Science denial robs the world of Christian voices in some of the most important issues of our time.

Not only does science denial thwart an evangelical presence in the sciences, but science denial damages the witness of evangelicals, and by association, Christianity in general.

The culture warriors tell us that evolution is causing atheism. Quite the opposite—the denial of evolution and other aspects of science is shipwrecking faith.

The Barna Group has told us for years that science denial is wrecking the faith of teens and young adults. Barna studied young adults who were regular churchgoers in their early teens but dropped out after age fifteen. The reasons for dropping out are varied and complex, but six themes consistently top the list. Always among the top six is the tense relationship between the church and science.[11]

The culture warriors tell us that evolution is causing atheism. Quite the opposite—the *denial* of evolution and other aspects of science is shipwrecking faith.

Stepping outside the church walls, we see damage to Christian witness in the community at large.

Over the last few years of reading and researching science denial during the pandemic, I saved and bookmarked countless anecdotes, social media posts, news reports, and polling data regarding the evangelical response to COVID. One poll, however, stopped me in my tracks and has haunted me since. White evangelicals are the least likely of all religious demographic groups to be vaccinated for COVID. Drilling down into the rationale behind vaccine refusal reveals a troubling trend. White evangelicals are also the religious demographic least likely to consider the health of their community in making a vaccination decision.[12]

Read that last sentence again and let it sink in.

Less than 50 percent of white evangelicals said they would consider the health of their community "a lot" when making a vaccination decision. Black Protestants and Catholics were much higher at 70 percent and 65 percent.

There is another religious demographic who likewise laps white evangelicals: atheists, agnostics, and those who identify as "nothing in particular." Sixty-eight percent of religiously unaffiliated Americans would consider the health of their community in making a vaccination decision.

Yale researchers wanted to know what, if anything, would reduce evangelical vaccine hesitancy. Maybe it was the messaging.[13] The researchers tried a message of reciprocity: Vaccination protects others in your community, who, in turn, protect you. They tried a message of responsibility: Wouldn't you feel bad if you infected someone else? They tried a message of values: Refusing a vaccine makes you reckless; don't you want to be brave and protect your community?

Nothing worked. Unvaccinated white evangelicals did not find any of these community-care, neighbor-loving messages persuasive.

Paying the Price

The most vulnerable among us are paying the highest price for evangelical science denial.

In middle-class America, where most of us live in clean neighborhoods and can pay our bills, we are often oblivious to the fallout of science denial.

Reaching herd immunity from a more transmissible virus requires a higher percentage of vaccinated people. When we're talking about a highly transmissible virus like COVID, the vaccination rate needed for herd immunity is 80–85 percent.

The fewer vaccinated people there are, the greater the playing field for viral replication. With more viral replication, more variants arise.

Americans living in the poorest counties died from COVID at nearly twice the rate of Americans living in wealthier counties.[14] This is not surprising, since families like the Zubias have fewer opportunities to work from home, socially isolate, and access health care.

Old Testament scholar Sandra Richter is an environmentalist, and she invites us to leave the comfort of clean homes and neighborhoods with plenty of clean water and see the devastation of mountaintop removal coal mining in Western Virginia and Eastern Kentucky, the despair of families in Madagascar because of deforestation, and the trauma caused by a lack of water, from Texas to India.[15]

Evangelical science denial up until now hasn't threatened public health. Science denial hasn't been life-threatening.

But it certainly is now.

Opposing evolution morphed into resisting COVID precautions and vaccinations and any attempt to address climate change. Wielding the same arguments against science and scientists used for decades in fighting evolution, evangelical science denial became immediately life-threatening in a pandemic. In the case of climate warming, science denial is life-threatening in the long term.

To what degree was the COVID pandemic prolonged because evangelicals vilified Anthony Fauci? How many people were exposed to the virus because pastors kept churches open and maskless? How many at-risk people died or were harmed by widescale vaccine refusal?

How many generations will pay the price of making global warming a battle in an evangelical culture war?

Fighting evolution perverted our approach to science. We are conditioned to distrust the experts. We are comfortable having our own facts and standing against the consensus. We listen to comforting, familiar voices, so we don't worry about it. God's in control. Faith over fear.

Our culture wars are killing people and wrecking faith.

We haven't made it easy to speak up. If we accept the evidence for evolution, we might be accused of rejecting the Bible.

If we suggest that wildfires and drought have a human-caused component, they'll think we are liberal, or worse.

Our culture wars are killing people and wrecking faith.

And if you're Beth Moore holding pastors and congregations to account for what they did and did not say and do during the COVID pandemic, you'll be told by Greg Locke with his ginormous evangelical megaphone to "sit down."[16]

Solution Aversion

The problem is not with evangelical response to need. Should a flood or tornado or hurricane strike, evangelical disaster relief organizations dispatch loaded eighteen-wheelers within hours. Texas Baptist Men and Churches of Christ Disaster Relief are just two among many responding domestically, while Samaritan's Purse and World Vision have global reaches.

Short-term solutions to global climate disasters are no problem. We've got you. Organizations like these, large and small, are absolutely needed as first responders to floods, hurricanes, tornadoes, and refugee crises.

The conversation changes, however, when we talk about long-term climate solutions. We are, as Katharine Hayhoe says, "solution averse."

We oppose regulations that would mitigate the climate-warming greenhouse gases that are exacerbating floods, hurricanes, tornadoes, and drought. We oppose the science because we oppose the regulations. Regulations mean more government. More government means encroachment on freedoms to do as we wish.

If we accept human-caused climate warming, we'd feel obligated to do something about it.

Accepting climate change and COVID science and evolution all come with a cost. All have solutions, so to speak, to which we are averse.

In human cultures, it is not so important to form beliefs that are true as it is to form beliefs that lead to social acceptance. What matters most, says Jonathan Rauch, is not what *I* believe but what *we* believe: "Once a belief becomes important to the way we think about ourselves or important to the group we identify with, changing it becomes very costly."[17]

In an evangelical culture where politics and faith converge and flow together, acceptance of science comes with a cost. Economic costs, curtailed freedoms, the disapproval of those in your church—all are possible.

In the case of evolution, there is a cognitive price to pay. Accepting the evidence for evolution means releasing long-held beliefs about biblical inerrancy. It means accepting that parts of the Bible can be truth without being literally, historically, and scientifically true.

The Mind of Christ

In Nathaniel Hawthorne's classic *The Scarlet Letter*, Hester Prynne is given, among other punishments, a lifelong sentence to wear a scarlet *A* for the crime of adultery. Yes, the crime. In Puritan America, gossiping, skipping church, swearing, or committing adultery did not simply land you in hot water with the local minister. Sins were crimes, and crimes were sins, and local courts meted out punishments.

For most of the last five hundred years, secular life and religious life were one and the same. A king ruled as God's right-hand man. Even if you were agnostic or an atheist, you dared not admit it in public. Such an admission could cost you your job, your family, or your status in the community.

Not so in the twenty-first century. Unbelief in God is a real option for anyone. We can easily navigate the modern world and never reference the supernatural.

Charles Taylor (*A Secular Age*) asks this important question: "Why was it virtually impossible not to believe in God, in say, 1500 in our Western society, while in 2000 many of us find this not only easy, but even inescapable?"[18] Citizens of the twenty-first century, says Taylor, see the natural world with its science and laws banishing the need for faith, belief, mystery, and religion. Modern science is the air we breathe.

As disparate as world religions are, people of all religious faiths agree on this one point: science doesn't have all the answers. What are followers of Jesus contributing to the conversation?

As Christians, we are called to truth. Speaking it. Defending it. Living it. Why be afraid of science? If God is truth, all truth is God's truth, including scientific truth.

Christians are called to have the mind of Christ. Noll describes this as a mind *for* Christ, thinking like a Christian across the spectrum of modern learning, from economics to history to the arts.[19] What does it mean to have a mind for Christ, specifically in a world of modern science? What, besides denial, are we adding to the conversation? Are there things that the Bible simply does not speak to? How can we approach these things with a mind for Christ?

Science and faith are not enemies. Science does not have all the answers.

With a mind for Christ, may we live as people of faith in a modern scientific world.

Discussion Prompts

Chapter 1

1. The "draw a scientist" activity is a popular start-of-the-year science lesson, from kindergarten to high school. The goal of the exercise is to reveal biases and stereotypes: What do students think scientists do? What kinds of people can be scientists? If you were to draw a scientist right now, what would you draw? Male? Female? White coat? Crazy hair? Stuff blowing up in a lab? Good guy or bad guy?
2. From sitcoms to cinema, popular culture implies that science and faith are at odds. In what ways does popular culture (movies, television, social media) perpetuate this message?
3. Have you (or someone you know) experienced a faith tradition in which science or scientists are suspect? What did that look like? Were all scientists suspect, or just some of them?
4. In a creationist narrative, all new scientific evidence is approached with foregone conclusions, says Caroline Matas, a scholar who studies American evangelicals. What are the ramifications of such an approach? Can you think of an example of a "foregone conclusion" someone might hold regarding an area of science?

5. Is creationism part of your story, either in the past or the present? If so, do you agree with Matas? Why or why not?
6. Evangelical Christians accept modern science, without question, in most areas—medicine, technology, air travel. Why might evolution, climate science, and pandemic science be suspect to many evangelicals?
7. In the Scopes monkey trial, William Jennings Bryan characterized evolution as unsupported by science and a danger to morality, society, and faith. Explain your understanding of evolution. What kinds of sources shaped your understanding?
8. Answers in Genesis, a young-earth creationist organization, interprets all scientific evidence through a "Bible lens"—any perceived conflict between science and the Bible is due to misinterpretation of the scientific evidence. Analyze this approach to scientific evidence. Why might such an approach be comforting? Why might such an approach be problematic?
9. If you are currently part of a faith community, how did your community respond to COVID precautions? Masking? Meeting in person? Vaccines? Did you see consistency of attitudes toward all precautions, or were some measures more accepted than others?
10. If you are currently part of a faith community, how did the leadership respond to the COVID pandemic? Did you have an opportunity to express your voice in the matter? Did you feel heard?
11. How did you see churches and other faith communities within your larger community responding to the COVID pandemic?
12. *Confirmation bias* is the tendency to seek out and accept only information or evidence that affirms our existing beliefs. Without realizing it, we may ignore informa-

tion that challenges our beliefs, or we may cherry-pick information we feel confirms our beliefs. Apply the concept of confirmation bias to modern science. What might that look like?

Chapter 2

1. The Kansas State Board of Education is one of many state and local boards who have found themselves in a tug of war between creationists and science educators. Are you aware of any such controversy with your local or state boards of education? The Kansas board initially jettisoned national science standards for standards composed by local creationists. If your local or state board did the same, what would your reaction be? How do you think your community would respond?
2. What are the doctrinal and social marks of evangelicalism listed in this chapter? Which, if any, would you reword? Which, if any, would you delete? Would you add to either list? Do these lists reflect your understanding of evangelicalism?
3. In the twenty-first century, being an evangelical does not mean identification with a specific religious tradition or denomination. Instead, being evangelical is a mindset that transcends groups. In your faith community, personal relationships, or in your community at large, can you identify aspects of an evangelical mindset? What does it look like?
4. Prior to reading this chapter, what did you know about the Moral Majority and the Christian Coalition? Were you aware of their origins in Republican politics? In your opinion, is this connection positive, negative, or neutral? Explain your thinking.

5. Conservatives' trust in science has fallen significantly since the 1970s. What factors do you think contributed to this decline?
6. For the sake of discussion, can you accept that evolution is the underlying foundation and organizing principle of modern biology? What is your comfort level? Explain your perspective.
7. "How evangelicals talk about the science of evolution influences how evangelicals talk about other areas of science." Do you agree or disagree with this statement? Explain your position.
8. Do you consider yourself a creationist? If so, how do you define *creationism*? If you do not consider yourself a creationist, describe creationism from the outside looking in. Have you ever participated in a creationist-sponsored event? If so, describe the event and your impressions of the event.
9. Are you familiar with any of the creationist organizations listed in this chapter? Spend a few minutes perusing some of the websites for these groups. Note your first impressions.
10. "Follow the science" became a catchphrase during the COVID pandemic. In what contexts did you see or hear the phrase used? Did you see consistency or inconsistency in meaning?
11. William Jennings Bryan, a folk hero known as the "Great Commoner," shaped evangelical perspectives on evolution for decades. In your opinion, what advantage did Bryan have over scientists in shaping perspectives about evolution?
12. Are all opinions equally valuable regarding scientific evidence? Why or why not?

Chapter 3

1. We trust our safety, our health, and so many other aspects of our lives to people, systems, and technology about which we have no direct knowledge. Airplane travel is an example—most of us will fly on an airplane without extensive training in aeronautics, airplane engineering, aerodynamics, or piloting. Why do we do so? How do you decide whom to trust? Which people do you believe? What institutions do you trust?
2. When a new scientific story hits the popular news outlets—television, radio, newsfeeds—how likely are you to dig deeper, to find out more about the topic? What sources will you likely turn to for more information? In your experience, do people judge an idea on the merits of the idea or the source of the idea? Give examples.
3. Logical fallacies are common patterns of reasoning which may seem valid, but are, in fact, flawed. "Appeal to authority" is one such fallacy: when we do not have direct knowledge about a topic, we appeal to those who speak with "authority"—those with advanced degrees, positions of power, or large platforms. Identify possible pitfalls of appealing to authority in science rather than to data or expertise.
4. The video of "America's Frontline Doctors" quickly went viral, as did several other videos during the COVID pandemic. These videos promoted views contrary to the consensus of epidemiologists, medical researchers, and physicians regarding viral risk, the need for precautions, and the efficacy of alternative cures. Did you watch any "alternative view" videos? What do you think is the appeal of such videos? Can you iden-

tify any common denominators in "alternative view" presentations?

5. "We'll just have to agree to disagree." Have you ever used this phrase in a conversation? Has it ever been said to you? In what context was the phrase used? Why do you think Tom Nichols calls "agree to disagree" a "conversational fire extinguisher"?
6. During the COVID pandemic, to reject the consensus of scientific experts was to be an independent thinker, one who doesn't simply go along with the crowd. "I did my own research" was a common refrain. When might "doing your own research" be appropriate? When might it be problematic?
7. "It's wise to have a cautious approach to authority and a robust respect for expertise." What might this look like in real life? Do you agree or disagree with the statement? Why?
8. Many classroom teachers attempt to avoid controversy with topics like evolution or climate change by telling students "You don't have to believe it, just learn it for the test." What message might this send to students? What does this statement say about the role of evidence in science?
9. Describe *science* in your own words. What are the defining characteristics of science? Define and describe *opinion*. Compare the two terms. "Science is a neutral tool"—do you agree or disagree with this statement? Defend your answer.
10. Briefly summarize things science *can't* do, according to this chapter. What would you add to the list?
11. How does the demand for absolute proof limit our understanding of science? What drives the demand for proof?

12. Smoking was good for you; then it wasn't. Cocaine was harmless in soft drinks; then it wasn't. There used to be nine planets; now there's just eight. How can we trust science when it flip-flops so often? Carl Zimmerman describes scientific papers as "status reports." Does the "never finished" aspect of science disturb you? Does it confuse you?

Chapter 4

1. Think about your own science education. What was the emphasis—scientific facts or the process of science? What are your scientific strengths? What are your weak areas? What do you wish you knew more about?
2. Explain *control* in a science experiment. Give examples of things that might be important to control.
3. What biases might researchers bring to a study? How might participants in a study be biased? How do we control for bias in experimenters and participants?
4. Summarize the peer review process. In your opinion, how important is peer review? Should researchers with advanced degrees or prestige be exempt from peer review? Defend your position.
5. Good researchers perform tests that could disprove their idea. Does this increase or decrease your confidence in the scientific process? Explain your answer.
6. How many violations of good research design can you identify in Andrew Wakefield's study? Which are the most egregious? Explain your choices.
7. Imagine that you are in the group gathered to hear "America's Frontline Doctors" on the steps of the Supreme Court. After hearing the presentation, you are given the microphone. What three questions do you ask?

8. Advocates for ivermectin as a treatment or preventative for COVID were quick to point out that it is a Nobel Prize-winning drug. Deconstruct the argument: Why was the Nobel Prize so important to ivermectin advocates? Can you identify possible flaws in the argument?
9. Ivermectin and hydroxychloroquine were popular alternative treatments throughout the COVID pandemic, despite not being approved by the FDA or endorsed by the consensus of infectious disease scientists. What explains the popularity of these alternative treatments?
10. Consider the popularity of America's Frontline Doctors and ivermectin and hydroxychloroquine as alternative COVID treatments. Identify areas of deficiency in scientific literacy that might explain their popularity.
11. Were you surprised to learn about the evangelical roots of America's Frontline Doctors? Why do you think America's Frontline Doctors especially resonated with evangelical Christians?
12. Sociologist Zeynep Tufekci says that reading an opposing viewpoint on social media is like hearing fans of an opposing team yell at us from the other side of the stadium—we retreat to our side with the people who think like we do. What does this look like when the topic is a poorly understood scientific issue? What are the risks of assessing fact-based scientific issues in an echo chamber of a like-minded group?

Chapter 5

1. Were you surprised to hear Dr. Kizzmekia Corbett's "It could have been faster" comment about the COVID vaccine? Explain her reasoning. Is Dr. Corbett's rationale convincing? Why or why not?

2. White evangelicals are by far the largest religious demographic rejecting the COVID vaccine. In fact, white evangelicals rejecting the vaccine outnumber atheists who reject the vaccine by a large margin. Suggest possible reasons for the gulf between white evangelical acceptance and atheist acceptance of the COVID vaccine.
3. Many evangelicals believe that scientists misrepresent the evidence for evolution. Is it reasonable to assume this mindset played a part in mistrust of scientists during the COVID pandemic? Why or why not?
4. The COVID pandemic was an unprecedented time. It seemed as if the news changed daily, if not hourly. Science communication was not always at its best. What was challenging for you as you tried to stay informed? What would have helped? What was frustrating?
5. Pseudoscience provides simplicity and certainty; science talks about the best evidence at the moment, not proof. Fad dietary supplements are often more popular than long-term fitness and healthy eating. Identify other examples of the certainty of pseudoscience versus the uncertainty of science.
6. Tell about a time you heard or read the phrase "faith over fear" during the pandemic. What was the context? Have you ever heard this phrase used outside the context of the COVID pandemic?
7. Evangelical response to the COVID pandemic often took the form of "God will protect me." Some, like Joy Pullmann, saw COVID precautions as a lack of trust in God and a sign of weakness in the modern church. Did you observe this response in your community, family, or faith community? If so, what did this response look like? Discuss this statement in the context of the pandemic: "It is God who decides when we die."

8. Read Matthew 4:5–7. Was "faith over fear" a top-of-the-temple moment for evangelicals? What parallels, if any, do you see between the story of Jesus at the top of the temple and "faith over fear" evangelicals in the pandemic?
9. Prominent evangelical leaders saw Satan at work in the pandemic—specifically, in the shuttering of churches. While some evangelicals adjusted to online services and found ways to serve their communities, others fought social distancing and mask mandates in court. Were you aware of churches acting in one of these two ways? If so, what was your impression at the time? Has your opinion changed?
10. Compass International, as well as numerous evangelical pastors, created and publicized forms and procedures for requesting religious exemptions to COVID vaccine mandates. Do you find it concerning that religious, nonmedical organizations and leaders are providing such guidance? Why or why not?
11. Over the last seventy years or so, various new scientific and technological advances have initially been feared as the book of Revelation's "mark of the beast." Recall the examples given in this chapter. Why might such conspiracies find a foothold?
12. Can a "faith over fear" attitude be both God-honoring and science-trusting? What would that look like?

Chapter 6

1. In evangelicalism, the intellect is not completely dismissed, but it is often devalued. This chapter cites examples from contemporary praise songs and well-known evangelical catchphrases. Do you agree or dis-

agree with this assessment? Why or why not? Can you think of other examples?

2. Historical biblical criticism set the stage for the divergence of evangelicals from generic Protestantism. Summarize *historical biblical criticism*. How does this approach to reading the Bible differ from a "plain reading" approach?
3. Evolution, together with historical biblical criticism, undermined the core of evangelicalism. Which aspects of evolution and biblical criticism were most offensive to evangelicals?
4. Separated from other Protestants, evangelicals built a massive empire of publication, broadcasting, and parachurch organizations. The subculture is so large, evangelical content producers never have to leave the bubble. What advantages might evangelicals find in this bubble? What are possible consequences of such a bubble?
5. Were you surprised to learn that Harvard, Princeton, and Yale all began as Christian universities? Why were these universities considered cautionary tales by founders of evangelical colleges?
6. Evangelical Bible colleges, schools of preaching, and most seminaries educate preachers, pastors, and leaders in a setting separate from the rest of academia. How important is "cross pollination" of theology with science, history, and the arts? Give examples to support your position.
7. Science departments at some evangelical universities, like the three mentioned in this chapter, teach evolution as the foundational principle of biology. Yet, in each of the three, faith informs education in biology. Is this approach inconsistent? Is it hypocritical? If you

were the chair of a biology department at a Christian university, how would you advise your faculty regarding the teaching of evolution?

8. What does the term *reverse snobbism* mean to you? Give examples. Have you read, heard, or experienced this in a religious context? In any context?
9. John MacArthur refers to scientists like Francis Collins as part of an elite, scholastic, scientific community that demands veneration. When we describe an athlete or an artist as elite, it is a compliment. But when "elite" is used synonymously with scholarship and science (as MacArthur does), the connotation is negative. Analyze the usage of the term *elite*—why is it sometimes positive, sometimes negative?
10. What is the backstory of *creation science*? Why was the term coined? Can creationism be studied using the scientific method? Why or why not?
11. Describe three possible responses by evangelicals working as research scientists or as academics. Identify the ramifications of each response.
12. In denying the evidence for evolution and the age of the earth, have evangelicals undermined their ability to understand the natural world? Discuss the possible fallout of telling evangelicals that biological and geological evidence can't be trusted.

Chapter 7

1. In a few areas of the world Christians are truly in danger, but most Christians, especially in the West, live without fear of prison or martyrdom. Yet, evangelicals often claim oppression. Jonathan Merritt says that evangelicalism is "predicated on an enemy to fight."

Is this your experience, whether from within evangelicalism or as an outside observer? If yes, what does this mindset look like?

2. Why are evangelicals drawn to an underdog story? What is attractive about the minority position?
3. Whether you accept evolution, reject evolution, or find yourself somewhere in the middle, think about the arguments you have heard against it. List as many as you can think of. How many are scientific arguments—evidence or studies refuting evolution? How many are arguments about cultural consequences of evolution?
4. Read through the "Wedge Document" (see the link in the endnotes). What are your impressions of the document? What are your thoughts regarding the Wedge's goal of overthrowing materialism (evolution science) and replacing it with a science that reflects Christian convictions?
5. *Worldview* is a euphemism often substituted for *culture* in evangelical circles. Christian private schools offer conferences and courses in "worldviews" and evangelical leaders talk about the need for a Christian worldview. What role does science play in worldview discussions? How is science portrayed as a combatant in culture wars?
6. Not all the residents of Glen Rose, Texas, accept an ancient age for the earth and evolution. Despite the evidence under their feet, many believe that humans and dinosaurs lived at the same time. What do you think leads to such a paradox?
7. Atheism is often cited as the hidden agenda of evolution. Is evolution a "get out of jail free" card for those who don't want to obey God? Is evolution "made up" by those who want to do as they please? Explain your reasoning.

8. "Evolution is agnostic, as is all science." Explain.
9. Why might evangelicals see politicians as allies in the fight against evolution?
10. Although it is not surprising that almost 50 percent of religious college students think evolution requires atheism, it is surprising that almost the same percentage of nonreligious college students believe the same. Suggest possible explanations for this finding.
11. Analyze this statement: "It is much easier to blame evolution for violence and atheism and even disrespectful children than it is to prove evolution false." Do you agree or disagree?
12. Evangelicals usually identify with the underdogs in the Bible. Standing against the majority feels noble and righteous. What if the majority is right? Explore this idea and journal a few sentences.

Chapter 8

1. Did you know the story of Mary Mallon before reading this chapter? What do you find most interesting about her story? If you had the chance to talk to Mary, how would you explain the situation to her?
2. In 1968, the United States Supreme Court ruled that laws forbidding the teaching of evolution are unconstitutional. During the following decades, creationism proponents brought various suits, all alleging violations of rights and freedoms. Briefly summarize the court cases. Which of the cases would you like to know more about? What questions would you ask?
3. How are "equal time" laws a call for rights and freedoms? Equal time laws are usually about evolution or climate science. Are groups fighting for equal time laws

for alternative views of gravity, the soil cycle, or cell theory? Why or why not?

4. Evangelical refusal to comply with COVID precautions is often couched in the language of religious liberty. How did the fight against evolution shape the anti-COVID response of evangelicals?
5. Going into the 2016 presidential election, evangelicals overwhelmingly believed that religious liberty was under attack. How did this mindset feed the politicization of the COVID pandemic? Give examples.
6. Under the guise of "obeying God rather than men," evangelical pastors fought to assemble in mask-free crowds. Some encouraged calling Thanksgiving gatherings a "funeral for a turkey." Read Acts 5:1–30. Compare the context for Peter's statement with evangelicals' use of the statement in the context of the pandemic.
7. Evangelicals merged baby Moses with the "I will not comply" meme, a meme often connected with gun rights and more recently with COVID restrictions. What were posters of the baby Moses version trying to say? What might you say to posters of this meme?
8. Both political pundits and evangelical leaders regularly cited violation of constitutional rights regarding COVID precautions. Assess the situation: To which law or laws should Christians appeal? Do circumstances matter?
9. The Great Barrington Declaration was broadly condemned as "ethically problematic." Describe some of the ethical problems with the plan.
10. In what ways are social distancing and working from home privileges not available to all? What should be the response of Christians to those without these privileges? Specifically, what would that response look like?

11. Curtis Chang observed, "Once Trump set the Republican culture down this path, he made it very difficult for Evangelical leadership to lead." What difficulties did pastors and leaders supportive of COVID precautions face? What difficulties did church members supportive of COVID precautions face when church leadership was not supportive?
12. As honestly as you can, appraise the impact of Christians during the COVID pandemic. What were Christians, particularly evangelical Christians, known for during the pandemic? If you are in a faith community, how does your assessment apply to your group?

Chapter 9

1. The circulation of blood may seem obvious, but it was not always known. Once circulation of blood was demonstrated, all sorts of doors opened leading to modern medicine. We learned, for example, that a drug injected intravenously would be distributed to the entire body. Live animals were used in the discovery of blood circulation and in many other discoveries. What are your thoughts regarding animal research prior to the twenty-first century? What are your thoughts about animal research in this century?
2. A *Washington Post* headline read "Only in our anti-truth hellscape could Anthony Fauci become a super-villain." How did a scientist with a stellar career become a supervillain in a matter of weeks? Opposition to Fauci was strong in evangelical circles. What fueled the fight?
3. What are the characteristics of a "secular" scientist, according to creationist resources? What are the characteristics of a "creation scientist"?

4. The film *Is Genesis History?* references the work of secular scientists regarding the age of the earth and evolution but adds "They're wrong." Since most working scientists are secular scientists (according to the film), what message does this send to viewers?
5. How would you respond to a documentary claiming that most of the world's scientists are wrong about cell theory? Or atomic theory? Or gravitational theory? What questions might you have for the producers of such a film?
6. Why do you think so much creationist attention is given to frauds like Piltdown man and *Archaeoraptor*? Were you surprised to learn that it was "secular" scientists who exposed the frauds?
7. If you were told that biologists and geologists purposely lie about evidence for an old earth or evolution, what might you conclude about scientists in general?
8. Textbooks published for Christian private and homeschools have a common theme: secular scientists have no proof, so they make things up. How likely is a worldwide cabal of unrelated scientists uniting to deceive people? What problems arise with a "They're all in on it" explanation for evidence?
9. The Creation Science Research Center extends the secular label beyond scientists to science educators. Why might educators find themselves in the crosshairs of creationists?
10. Evangelicals love a sports hero with a faith story—hence the popularity of the "I Am Second" stories. Enthusiastic evangelical scientists like Francis Collins, however, don't share the same adulation. What explains the discrepancy?
11. What does the phrase "shooting the messenger" mean? How is the phrase applicable in the context of evo-

lution, COVID science, and climate science? What motivates such a response?

12. Is there a relationship between how we characterize the messenger and how we receive the message? Can we ask honest questions about science without vilifying scientists? What might that look like?

Chapter 10

1. Explain the difference between climate and weather. Give an example of each.
2. What is the function of carbon dioxide in the atmosphere? What kinds of things happen when there is too much carbon dioxide in the atmosphere?
3. Why is global warming considered a threat multiplier? What threats are exacerbated by higher global temperatures? How are vulnerable populations impacted?
4. Senator Inhofe threw a snowball in the Senate chamber as "proof" that global warming is a hoax. Dissect his demonstration. Does a snowball in February debunk climate change? Explain your answer.
5. Why do young-earth creationists have a problem with climate-change science? To what do creationists attribute climate changes?
6. Many Christians disassociate the biblical command to "steward the earth" from human-caused global warming. What motivates this disassociation? What does this disassociation look like in real life?
7. The earth is a "disposable planet," according to John MacArthur. How would you respond to MacArthur? What questions would you ask him?
8. How did climate-change science become political? What are the primary concerns of those who identify as politically conservative or libertarian?

9. In your opinion, are most people who deny human-caused global warming rejecting the scientific evidence or are they rejecting the solutions needed to address the problem? Explain.
10. Climate scientists have been accused of the same offenses as evolution scientists and COVID scientists: hiding evidence, manipulating data, having a political agenda. These are three very different areas of science—why the overlap in criticisms?
11. Creationist think tanks like the Discovery Institute also provide guidance for fighting climate science in schools. What similarities do you see in the "both sides" and "strengths and weaknesses" approaches to teaching evolution and climate science?
12. Science denial regarding evolution, COVID, and human-caused global warming often focuses on "proof," as in "I won't believe it unless I have absolute proof." What issues arise from the demand for absolute proof in science? Is absolute proof achievable? Why or why not?

Chapter 11

1. What are embryonic stem cells? Why do we call an embryonic cell a *stem* cell? Embryonic stem cells used in research are not retrieved directly from an embryo—explain.
2. What are the two types of adult stem cells? Describe each type.
3. Describe the advantages of using stem cells to test new drugs. Why do stem cells often give us a better picture of a drug's actions than lab animals can give us?
4. Fetal tissues used in research are not directly retrieved from a fetus—explain.
5. During the COVID pandemic, were you aware of vaccine opposition related to fetal cell concerns? If so, what

did you hear or read about this concern? Was use of fetal cells a factor in your vaccination decision? Explain.

6. Summarize Curtis Chang's analogy of the transcontinental railroad. How are vaccines and medicines developed with embryonic and fetal tissues like the transcontinental railroad?
7. In your opinion, is the use of fetal tissues to develop lifesaving medicines consistent with support of pro-life positions? State your case.
8. Victoria Gray was born with a devastating genetic disorder, sickle cell disease. How was gene editing used to treat Victoria? Do you support the use of gene editing to treat existent disease? Why or why not?
9. Is it important to you to hear from people of faith who also have expertise in areas of biotechnology? Why or why not? What perspectives might such experts add?
10. Are some (or all) of the biotechnologies described in this chapter scary to you? If one of these procedures could restore the health or save the life of a loved one, would that change your feelings? Describe your thoughts.
11. What kinds of regulations would you like to see regarding these technologies: stem cells, fetal cells, gene editing, mitochondrial transplants, and genetically modified food crops?
12. How might science denial and mistrust of scientists in the contexts of evolution, COVID, and climate influence attitudes toward biotechnology?

Chapter 12

1. How might science inform religious beliefs? How might faith inform implementation of scientific innovations? What could people of faith add to the discussion?

2. Describe the ethical issues involved in He Jiankui's gene editing in human embryos.
3. What does the phrase "stay in your lane" mean to you? How is Gould's philosophy of nonoverlapping magisteria a fancy way of saying "stay in your lane"? Can science and faith share a "lane" without denying either science or faith?
4. Has fighting evolution shaped the way evangelicals approach science? Can you think of a real-life example?
5. Why is seeking information within our confirmation biases so easy to do? What are the costs of looking outside our biases?
6. If we accept the evidence for human evolution instead of fighting it, what could people of faith add to the conversation? If we accept the evidence for human-caused climate change instead of fighting it, what could people of faith add to the conversation?
7. A consequence of science denial is the hesitance of young evangelicals to be scientists, especially in the fields of evolutionary biology and climate. What is the cost to evangelicals? What is the cost to the wider community?
8. White evangelicals lag behind nonreligious people in considering the well-being of the community when making a vaccine decision. Do you think a culture of science denial contributes to this response? Why or why not?
9. Sometimes it is not easy to speak up for science in evangelical circles. What might the cost be to an evangelical who accepts evolution? What might be the cost to an evangelical who accepts human-caused climate change? How might these costs be mitigated?
10. Evaluate this statement: What *I* believe is not as important for social acceptance as is what *we* believe. Do you agree or disagree?

11. Charles Taylor says that it is possible, even easy, to live in modern societies without professing faith. Do you see this in your local community? Do you see this in the wider community? In the past, it was not easy to go public about nonbelief. What changed?
12. Science does not have all the answers in life. What does it mean to be people of faith in a modern scientific world?

Notes

Chapter 1

1. Janet Kellogg Ray, *Baby Dinosaurs on the Ark? The Bible and Modern Science and the Trouble of Making It All Fit* (Grand Rapids: Eerdmans, 2021).

2. Private group text discussion board with author, March 2, 2021.

3. David Croom (@dailycallout), "The same people who invented the vaccine believe humans evolved from apes!" Twitter, September 8, 2021, https://tinyurl.com/2panupjr.

4. Jessie Szalay, "Scopes Monkey Trial: Science on the Stand," Live Science, September 30, 2016, https://tinyurl.com/2nxnbvmn.

5. Here's an image from the local newspaper promoting Monkey Sunday: "J. Frank Norris: One Foot in the Pulpit, One Foot in the Witness Stand," *Hometown by Handlebar* (blog), July 17, 2022, https://tinyurl.com/2p8da9kr.

6. Sugar Hill Church, "Herschel Walker Interview," YouTube video, 35:05, March 13, 2022, https://tinyurl.com/ywrds5mr.

7. Caroline Matas, "Welcome to the War: What a Creationist Conference Can Teach Us about Evangelical Vaccine Resistance," Religion Dispatches, October 11, 2021, https://tinyurl.com/2p96u84u.

8. Edward Humes, *Monkey Girl: Evolution, Education, Religion, and the Battle for America's Soul* (New York: HarperCollins, 2007), 144.

Chapter 2

1. Jessica Johnson Webb, "Kansas Is a Fossil Hunter's Paradise—Here's How and Where to Explore It," *Roadtrippers*, April 2, 2021, https://tinyurl.com/3dcmh545.

2. Edward Humes, *Monkey Girl: Evolution, Education, Religion, and the Battle for America's Soul* (New York: HarperCollins, 2007), 146–54.

3. "Geology Removed from Kansas History Book," United Press International, August 21, 1999, https://tinyurl.com/bdhe5y2m.

4. Humes, *Monkey Girl*, 154–56.

5. Timothy Keller, "The Decline and Renewal of the American Church: Part 2—The Decline of Evangelicalism," Gospel in Life, Winter, 2022, https://tinyurl.com/3zf2fwnw; and John Green, Mark Noll, and Richard Cizik, "Evangelicals v. Mainline Protestants," Frontline, April 29, 2004, https://tinyurl.com/2was4d9n.

6. Interview with Mark Noll in Green, Noll, and Cizik, "Evangelicals v. Mainline Protestants."

7. Ryan Burge, "Are We All Evangelicals Now? How the Term Has Grown to Blur Theology and Ideology," Religion Unplugged, March 11, 2021, https://tinyurl.com/35y8y9u2; and Burge, "Why 'Evangelical' Is Becoming Another Word for 'Republican,'" *New York Times*, October 26, 2021, https://tinyurl.com/ykzc6mnu.

8. Gregory A. Smith, "More White Americans Adopted Than Shed Evangelical Label during Trump Presidency, Es-

pecially His Supporters," Pew Research Center, September 15, 2021, https://tinyurl.com/2h2sd55b.

9. Ryan Burge, "Another Word for 'Republican.'"

10. Daniel K. Williams, "White Southern Evangelicals Are Leaving the Church," *Christianity Today*, August 2, 2022, https://tinyurl.com/4kytdtjv.

11. Emma Green and Julia Longoria, "'Evangelical' Is Not a Religious Identity. It's a Political One," May 13, 2021, in *The Experiment*, produced by Katherine Wells and Alvin Melathe, podcast, 38:53, https://tinyurl.com/2jrcp3d6.

12. Green and Longoria, "'Evangelical' Is Not a Religious Identity."

13. Susan Davis, "This Conservative Leader Is Trying to Make White Evangelical Politics Less White," NPR, July 5, 2022, https://tinyurl.com/3ujzsxy8.

14. Dhruv Khullar, "Faith, Science, and Francis Collins," *New Yorker*, April 7, 2022, https://tinyurl.com/2p85tu26; and Scott Jaschik, "Conservative Distrust of Science," *Inside Higher Ed*, March 29, 2012, https://tinyurl.com/bde9jn8e.

15. Jaschik, "Conservative Distrust of Science."

16. I wrote about the evidence for evolution in *Baby Dinosaurs on the Ark? The Bible and Modern Science and the Trouble of Making It All Fit* (Grand Rapids: Eerdmans, 2021).

17. "The Biology Wars: The Religion, Science and Education Controversy," Pew Research Center, December 5, 2005, https://tinyurl.com/4j8bxydd.

18. Marc Fisher, "'Follow the Science': As the Third Year of the Pandemic Begins, a Simple Slogan Becomes a Political Weapon," *Washington Post*, February 11, 2022, https://tinyurl.com/2p92bmxn.

19. Rand Paul (@RandPaul), "Open the schools! I wonder where the 'listen to the science' people will go when the sci-

ence doesn't support their fearmongering or their chosen narrative?" Twitter, October 9, 2020, https://tinyurl.com/5ev32zt6.

Chapter 3

1. Karen Freifeld, "Judge Tosses Houston Hospital Workers' Lawsuit over Vaccine Requirement," Reuters, June 14, 2021, https://tinyurl.com/5n7rkfnt.

2. Michelle Charles, "City of Stillwater Drops Mask Requirement after Businesses Threatened," *Stillwater News Press*, May 1, 2020, https://tinyurl.com/ynr9w936.

3. Rod Dreher, "Mask Truthers," *American Conservative*, May 2, 2020, https://tinyurl.com/3wk6ha72.

4. See more about groupthink during the pandemic in this podcast interview with David French and Curtis Chang: "Christians and the Vaccine," BioLogos, originally aired March 18, 2021, https://tinyurl.com/28s88t3n.

5. "Nobel Prize in Quackpottery: Chemistry," *Guardian*, October 12, 2012, https://tinyurl.com/2tvxwsds.

6. John Burnett, "The Number of Americans Who Say They Won't Get a COVID Shot Hasn't Budged in a Year," NPR, May 10, 2022, https://tinyurl.com/bdhfwdd4.

7. Tom Nichols, *The Death of Expertise: The Campaign against Established Knowledge and Why It Matters* (New York: Oxford University Press, 2017), iii.

8. Michael B. Berkman and Eric Plutzer, "Defeating Creationism in the Courtroom, but Not in the Classroom," *Science*, January 28, 2011, https://tinyurl.com/2p8na77r.

9. Berkman and Plutzer, "Defeating Creationism in the Courtroom."

10. Adelle M. Banks, "Francis Collins on COVID-19 Politics: 'The Culture War Is Literally Killing People,'" Religion News Service, February 3, 2020, https://tinyurl.com/2s3f5srn.

11. "Why Are Americans Putting Up with This?," *Tucker*

Carlson Tonight, Fox News, aired January 10, 2022. Read the transcript here: https://tinyurl.com/2p8d7bsa.

12. Brian Thomas, "Do We Always Believe What Scientists Say?," Institute for Creation Research, July 16, 2014, https://tinyurl.com/3et8pkjw.

13. Thomas, "Do We Always Believe?"

14. Thomas G. Safford, Emily H. Whitmore, and Lawrence C. Hamilton, "Follow the Scientists? How Beliefs about the Practice of Science Shaped COVID-19 Views," *Journal of Science Communication* 20, no. 7 (2021): A03, https://tinyurl.com/384ahv67.

15. See interesting insights from science historian Michael Gordin about the "valorization of expertise": Marc Fisher, "'Follow the Science': As the Third Year of the Pandemic Begins, a Simple Slogan Becomes a Political Weapon," *Washington Post*, February 11, 2022, https://tinyurl.com/2p92bmxn.

16. Thomas, "Do We Always Believe?"

17. *Parks and Recreation*, season 6, episode 5, "Gin It Up!," directed by Jorma Taccone, written by Greg Daniels, Michael Schur, and Matt Murray, aired October 17, 2013, on NBC.

18. Kathleen Hall Jamieson, "How to Debunk Misinformation about COVID, Vaccines and Masks," *Scientific American*, April 1, 2021, https://tinyurl.com/5t56xhh2.

19. Carl Zimmer, "How You Should Read Coronavirus Studies, or Any Science Paper," *New York Times*, June 1, 2020, https://tinyurl.com/57d7kk8k.

Chapter 4

1. Gardiner Harris, "Journal Retracts 1998 Paper Linking Autism to Vaccines," *New York Times*, February 2, 2010, https://tinyurl.com/bdf4fr7t.

2. "Measles Outbreak—California, December 2014–Feb-

ruary 2015," Centers for Disease Control and Prevention, February 20, 2015, https://tinyurl.com/yc8akneu.

3. Jacqueline Howard, "Minnesota Measles Outbreak Exceeds Last Year's Nationwide Numbers," CNN Health, June 2, 2017, https://tinyurl.com/2p8fey79; and Faiza Mahamud and Glenn Howatt, "In Measles Outbreak, a Misconception about Vaccines Still Plagues Somali Community," *Star Tribune*, April 22, 2017, https://tinyurl.com/3y8kwkdk.

4. In 1989, two chemists announced that they had achieved "cold" fusion—a monumental announcement, but all was not as it appeared. Read more about it in this case study: "Cold Fusion: A Case Study for Scientific Behavior," Understanding Science, University of California Museum of Paleontology, Berkeley, and the Regents of the University of California, 2012, https://tinyurl.com/39xutr7v.

5. Even Nobel Prize winners are subject to peer review. A Prize-winning chemist had to retract a paper (not the Nobel-winning paper) because of errors in her data: Bruce Y. Lee, "Nobel Prize Winner Frances Arnold Retracts Paper, Here Is the Reaction," *Forbes*, January 5, 2020, https://tinyurl.com/4rf54dma.

6. Julia Belluz, "The Research Linking Autism to Vaccines Is Even More Bogus Than You Think," Vox, updated January 10, 2017, https://tinyurl.com/4ah6jsad; and Fiona Godlee, Jane Smith, and Harvey Marcovitch, "Wakefield's Article Linking MMR Vaccine and Autism Was Fraudulent," *BMJ*, January 6, 2011, https://tinyurl.com/79ducs9j.

7. Sam Shead, "Facebook, Twitter and YouTube Pull 'False' Coronavirus Video after It Goes Viral," CNBC, July 28, 2020, https://tinyurl.com/3afh9xye.

8. Here's the transcript for the press conference: "America's Frontline Doctors SCOTUS Press Conference Transcript," Rev, July 27, 2020, https://tinyurl.com/45nbx6cp.

9. "Doctors SCOTUS Press Conference."

10. "Doctors SCOTUS Press Conference."

11. "Doctors SCOTUS Press Conference."

12. Vanessa Romo, "Poison Control Centers Are Fielding a Surge of Ivermectin Overdose Calls," NPR, September 4, 2021, https://tinyurl.com/52e67j33.

13. Jonathan Jarry, "The Ivermectin Train Cannot Stop," Office for Science and Society, McGill University, October 30, 2021, https://tinyurl.com/mw69kdb2; and Sara Reardon, "Flawed Ivermectin Preprint Highlights Challenges of COVID Drug Studies," *Nature* 596 (2021): 173–74, https://tinyurl.com/46njdmmu.

14. Gilmar Reis et al., "Effect of Early Treatment with Ivermectin among Patients with Covid-19," *New England Journal of Medicine* 386 (2022): 1721–31, https://tinyurl.com/bp9ev65v; and Hassan Yahaya, "Not a Miracle Drug: Experts on Ivermectin in COVID-19 Treatment," *Medical News Today*, March 11, 2022, https://tinyurl.com/yj37wvuv.

15. Adam Rogers, "The Strange and Twisted Tale of Hydroxychloroquine," *Wired*, November 11, 2020, https://tinyurl.com/2p955khp.

16. Keith Rabois, "Randomized controls are horrible ideas. Largest impediment to progress in health spans," Twitter, March 24, 2020, https://tinyurl.com/uxy4mjz2, quoted in Rogers, "Twisted Tale of Hydroxychloroquine."

17. Kai Kupferschmidt, "Three Big Studies Dim Hopes That Hydroxychloroquine Can Treat or Prevent COVID-19," *Science*, June 9, 2020, https://tinyurl.com/mw7cs34f; "Hydroxychloroquine Doesn't Benefit Hospitalized COVID-19 Patients," National Institutes of Health, November 24, 2020, https://tinyurl.com/bdfwympa; Caleb P. Skipper et al., "Hydroxychloroquine in Nonhospitalized Adults with Early

COVID-19," *Annals of Internal Medicine* 173, no. 8 (2020): 623–31, https://tinyurl.com/5n7hj5ax.

18. Geoff Brumfiel, "This Doctor Spread False Information about Covid. She Still Kept Her Medical License," NPR, September 14, 2021, https://tinyurl.com/2p8vcfxp.

19. Julia Duin, "Pro-Vax Christian Facebook Group Seeks Refuge from Friends, Family and Faith Communities," *Newsweek*, October 8, 2021, https://tinyurl.com/y4d6bprb.

20. David D. Kirkpatrick, "The 2004 Campaign: The Conservatives; Club of the Most Powerful Gathers in Strictest Privacy," *New York Times*, August 28, 2004, https://tinyurl.com/2p8vumt5; Robert O'Harrow Jr., "God, Trump and the Closed-Door World of a Major Conservative Group," *Washington Post*, October 25, 2021, https://tinyurl.com/y5yzeu6w; Anne Nelson, "Anatomy of Deceit: Team Trump Deploys Doctors with Dubious Qualifications to Push Fake Cure for Covid-19," *Washington Spectator*, September 20, 2020, https://tinyurl.com/2k24yw3t; and Nelson, "Holding Democracy in the U.S. Hostage," *Washington Spectator*, October 10, 2019, https://tinyurl.com/3rpwy5mc.

21. Heidi St. John, during an interview with Simone Gold, *Off the Bench with Heidi St. John*, podcast audio, November 12, 2020, https://tinyurl.com/54hubupb.

22. Marcus Lamb, "Dr. Simone Gold was fired from her job as an Emergency Room physician after appearing on Capitol Hill with America's Frontline Doctors," Facebook, August 14, 2020, https://tinyurl.com/mryf8ccd.

23. Marcus Lamb, "From the Desk of Marcus Lamb: March 2021," *Empowered by the Spirit* (blog), March 12, 2021, https://tinyurl.com/4cfyjxp5.

24. Zeynep Tufekci, "How Social Media Took Us from Tahrir Square to Donald Trump," *MIT Technology Review*, August 14, 2018, https://tinyurl.com/3hrj76b2.

Chapter 5

1. Patrick Wascovich, "Scientist at Forefront of COVID-19 Vaccine Development Shares Insights," *Center Times Plus*, March 2, 2021, https://tinyurl.com/5f5dzryc.

2. Wascovich, "Scientist at Forefront of COVID-19 Vaccine Development."

3. Courtland Milloy, "Kizzmekia Corbett Is Working on a Vaccine for COVID-19. She's Using both Science and Faith to Do So," *Washington Post*, August 19, 2020, https://tinyurl.com/29wn3ze2.

4. Yonat Shimron, "Black Protestants Aren't Least Likely to Get a Vaccine; White Evangelicals Are," Religion News Service, March 5, 2021, https://tinyurl.com/yckru4m9.

5. Cary Funk and John Gramlich, "10 Facts about Americans and Coronavirus Vaccines," Pew Research Center, September 20, 2021, https://tinyurl.com/3jt86kvr.

6. Centers for Disease Control and Prevention, "COVID-19 Incidence and Death Rates among Unvaccinated and Fully Vaccinated Adults with and without Booster Doses during Periods of Delta and Omicron Variant Emergence — 25 U.S. Jurisdictions, April 4–December 25, 2021," *Morbidity and Mortality Weekly Report* 71, no. 4 (2022): 132–38, January 28, 2022, https://tinyurl.com/3z9hwx2v.

7. Henry Morris, "The Vanishing Case for Evolution," *Acts & Facts*, February 1, 2009, https://tinyurl.com/ycksu2p4.

8. Allan Smith, "Conservative Hostility to Biden Vaccine Push Surges with Covid Cases on the Rise," NBC News, July 19, 2021, https://tinyurl.com/3xwskz9h.

9. Eric Bradner, "5 Takeaways from CPAC's Summer Gathering in Texas," CNN, July 12, 2021, https://tinyurl.com/2p92n7m2.

10. Rand Paul (@RandPaul), "We are at a moment of truth

and a crossroads," Twitter, August 8, 2021, https://tinyurl.com/2hxcfc72.

11. Joy Pullmann, "For Christians, Dying from COVID (or Anything Else) Is a Good Thing," *Federalist*, October 18, 2021, https://tinyurl.com/pwjrd4nf.

12. Bill Perkins, "Bill Gates' Top Virologist Says 'Don't Take the COVID Vaccine!'" Compass International, March 19, 2021, https://tinyurl.com/4hy74y39.

13. Nate Flannagan, "Pastor Dies of Coronavirus Days after Saying 'My Faith Has Never Wavered,'" *Christian Today*, March 31, 2020, https://tinyurl.com/4dsbm3de.

14. Aleem Maqbool, "Coronavirus: Pastor Who Decried 'Hysteria' Dies after Attending Mardi Gras," *BBC News*, April 6, 2020, https://tinyurl.com/5atymk2d.

15. James Crowley, "Conservative Pastor Who Said 'I'm Not Going to Get the Coronavirus' Gets the Coronavirus," *Newsweek*, November 17, 2020, https://tinyurl.com/46cjsey4.

16. Alex Woodward, "A Phantom Plague: America's Bible Belt Played Down the Pandemic and Even Cashed In. Now Dozens of Pastors Are Dead," *Independent*, April 24, 2020, https://tinyurl.com/bde2pnjt.

17. Owen Strachan, "John MacArthur's Epic Stand," *To Reenchant the World* (blog), January 12, 2022, https://tinyurl.com/y3u8mjn3.

18. Strachan, "John MacArthur's Epic Stand."

19. Strachan, "John MacArthur's Epic Stand."

20. Oliver Milman, "Co-Founder of Christian TV Network That Railed against Vaccines Dies of Covid-19," *Guardian*, December 1, 2021, https://tinyurl.com/4p5nybur.

21. Carma Hassan, "Christian Television Network Founder and Preacher Marcus Lamb, Who Discouraged Vaccinations, Dies after Being Hospitalized for COVID-19," CNN Business, December 2, 2021, https://tinyurl.com/yc433vf5.

22. Bill Perkins, "Known Cures/Treatments for Covid-19," Compass International, February 11, 2021, https://tinyurl.com/2rn2p2dt.

23. Ian Lovett, "White Evangelicals Resist Covid-19 Vaccine Most among Religious Groups," *Wall Street Journal*, July 28, 2021, https://tinyurl.com/ysjpn8rp.

24. Bill Perkins, "Religious Exemptions Available to Avoid Employer-Mandated Covid-19 Vaccine," Compass International, August 5, 2021, https://tinyurl.com/mr3a8dy6.

25. John Gideon Hartnett, "Moderna Vaccine May Create Transhumans," *Bible Science Forum* (blog), September 3, 2020, https://tinyurl.com/2p8982s7.

26. Joe Schwarcz, "Madej Madness," Office for Science and Society, McGill University, November 17, 2021, https://tinyurl.com/bdhvyrve.

27. Jack Goodman and Flora Carmichael, "Coronavirus: False and Misleading Claims about Vaccines Debunked," *BBC News*, July 26, 2020, https://tinyurl.com/e8pdtm4z. See also "False Claim: A COVID-19 Vaccine Will Genetically Modify Humans," Reuters, May 18, 2020, https://tinyurl.cm/yckem6rf.

28. Mary Louise Kelly, "In the '24th Mile' of a Marathon, Fauci and Collins Reflect on Their Pandemic Year," NPR, March 9, 2021, https://tinyurl.com/msmp7nf2.

29. Mark Wingfield, "Francis Collins: 'Give God the Glory' for Vaccines 'but Roll Up Your Sleeve,'" Baptist News Global, July 23, 2021, https://tinyurl.com/5n97ftsy.

30. Curtis Chang is a theologian and former pastor who created a series of videos regarding common questions Christians have about the COVID vaccine. This video addresses the mark of the beast: "Is the COVID Vaccine the 'Mark of the Beast'?," Christians and the Vaccine, March 6, 2021, video, 18:08, https://tinyurl.com/24vef38a.

Chapter 6

1. John Piper, "Should Pastors Have to Go through a Certain Amount of Schooling?," *Desiring God* (blog), March 30, 2009, https://tinyurl.com/5a3hwa6t.

2. Joel Houston and Matt Crocker, "Tear Down the Walls," sung by Hillsong United on *Across the Earth* (Integrity Music), 2009, track 7, accessed August 8, 2022, https://tinyurl.com/3vcef8x4.

3. Brooke Ligertwood, "New Wine," sung by Hillsong Worship on *There Is More* (Hillsong Music Publishing), 2017, accessed August 8, 2022, https://tinyurl.com/2nak5z93.

4. Here's a good interview with John Green comparing mainline Protestantism and evangelicalism: John Green, interview by *Frontline*, PBS, December 5, 2003, https://tinyurl.com/y9rzp2mm.

5. Read more about it: Pete Enns interviews Dr. Christopher M. Hays about his book *Evangelical Faith and the Challenge of Historical Criticism*, *The Bible for Normal People* (blog), 2013, https://tinyurl.com/2p987hzm.

6. International Council on Biblical Inerrancy, *The Chicago Statement on Biblical Inerrancy*, 1978, https://tinyurl.com/47r4xpnz.

7. Denny Burk (@DennyBurk), "Do you affirm the doctrine of inerrancy as articulated in the Chicago Statement?" Twitter, January 24, 2022, https://tinyurl.com/ytawcpe2.

8. Molly Worthen, *Apostles of Reason: The Crisis of Authority in American Evangelicalism* (New York: Oxford University Press, 2014), 23.

9. Nathan O. Hatch and Michael S. Hamilton, "Can Evangelicalism Survive Its Success?," *Christianity Today*, October 5, 1992, 20–31.

10. Josh Buice, "Why Your Pastor Should Say 'No More' to

Beth Moore," *Delivered by Grace* (blog), May 24, 2016, https://tinyurl.com/4jyxsrem.

11. David Cloud, "Bob Jones University Past and Present," Way of Life Literature, December 15, 2020, https://tinyurl.com/4cs8stwh.

12. Ken Ham, "Choosing the Right Christian College," Answers in Genesis, October 2, 2018, https://tinyurl.com/huk4aj6e.

13. "Online Self-Paced Course Syllabus GSU-1132 Life Sciences," Moody Bible Institute, Chicago, Illinois, accessed November 26, 2022, https://tinyurl.com/5n96r3fp.

14. Matthew Emile Vaughn, "On Schools of Preaching," *Journal of Faith and the Academy* 5, no. 1 (2012): 60–74.

15. Mark Noll, *The Scandal of the Evangelical Mind* (Grand Rapids: Eerdmans, 1994), 19.

16. Stephen DeRose, "The Top 10 Evangelical Seminaries in the U.S.," Successful Student, May 22, 2022, updated September 29, 2022, https://tinyurl.com/mswnjkte.

17. Perry L. Glanzer, "What Is the Difference between a Christian College and a Christian University?," *Christian Scholar's Review*, August 14, 2020, https://tinyurl.com/38znnw9k.

18. Noll, *Scandal of the Evangelical Mind*, 19.

19. "Mission and Core Values," Department of Biology, Baylor University, Waco, Texas, accessed November 26, 2022, https://tinyurl.com/yckjreks.

20. "Mission and Core Values," Department of Biology, Baylor University.

21. "About the Department," Department of Biology, Abilene Christian University, Abilene, Texas, accessed January 4, 2023, https://tinyurl.com/5a9pmcb6.

22. "Calvin College Biology Department Perspectives on Evolution," Calvin University, February 4, 2011, https://tinyurl.com/bdhcesd4.

23. Timothy Keller, "The Decline and Renewal of the

American Church: Part 2—The Decline of Evangelicalism," *Gospel in Life*, Winter, 2022, https://tinyurl.com/3zf2fwnw.

24. Eugenie C. Scott, *Evolution vs. Creationism: An Introduction* (Berkeley: University of California Press, 2009), 26.

25. Russ Miller, "Defending Biblical Creation" (presentation, Steeling the Mind conference, Coeur d'Alene, ID, February 16, 2019).

26. John MacArthur, "Evangelicals, Evolution, and the BioLogos Disaster," interview with Phil Johnson, Grace to You, https://tinyurl.com/3tmzk8c6.

27. Ronald L. Numbers, *The Creationists: From Scientific Creationism to Intelligent Design* (Cambridge, MA: Harvard University Press, 2006), 269.

28. "The Biology Wars: The Religion, Science and Education Controversy," Pew Research Center, December 5, 2005, https://tinyurl.com/4j8bxydd.

29. Noll, *Scandal of the Evangelical Mind*, 177–78.

30. Noll, *Scandal of the Evangelical Mind*, 11.

Chapter 7

1. Jonathan Merritt, interview by Jen Hatmaker, *For the Love* podcast audio, March 1, 2022, https://tinyurl.com/ycxk8wba.

2. Greg Hall, "Welcome to the War," Answers in Genesis, September 10, 2016, https://tinyurl.com/yh2tp6k3.

3. Molly Worthen, *Apostles of Reason: The Crisis of Authority in American Evangelicalism* (New York: Oxford University Press, 2014), 224.

4. Ronald L. Numbers, *The Creationists* (Cambridge, MA: Harvard University Press, 1992), 314.

5. *Congressional Record*, June 16, 1999 (statement of Rep. Tom DeLay), https://tinyurl.com/24dzv996.

6. Kenneth B. Cumming, "Review of the PBS 'Evolution'

Series," Institute for Creation Research, December 1, 2001, https://tinyurl.com/52fcbb7b.

7. Discovery Institute, *The "Wedge Document": "So What?,"* 1999, published April 2019, https://tinyurl.com/3r8z2d9w.

8. Center for the Renewal of Science and Culture, "The Wedge Document," National Center for Science Education, October 14, 2008, https://tinyurl.com/3bppfxc8.

9. Jeffrey McMurray, "Scientists Study Foes' Ways at Creation Museum," NBC News, June 26, 2009, https://tinyurl.com/2p8uav4n; see A. Charles Ware, *"It Doesn't Take a Ph.D.! The Cure for a Culture in Crisis,"* Answers in Genesis, directed by Paul Varnum, 2004, on YouTube, 44:47, posted April 4, 2021, https://tinyurl.com/48re69b4.

10. "Truth Matters Conference," Grace to You (Williamstown, KY, May 18–20, 2022), https://tinyurl.com/yckup6mz.

11. Owen Strachan (@ostrachan), "Young pastors and theologians: stop doing theology by a hermeneutic of shame. Stop being embarrassed by the Bible," Twitter, May 21, 2022, https://tinyurl.com/3etumtmm.

12. Robyn Ross, "Tracking Creation in Glen Rose," *Texas Observer*, April 4, 2012, https://tinyurl.com/5f9963dz.

13. The Clergy Letter Project, https://tinyurl.com/2p87p5ke.

14. Ken Ham, "1,000 Scientists Sign Up to Dissent from Darwin," Answers in Genesis, February 11, 2019, https://tinyurl.com/49894s4b.

15. John MacArthur, "Is Evolution Compatible with Christianity?," Grace to You, August 28, 2009, https://tinyurl.com/3x6a87nh.

16. Eric Hovind, interview with Brian Norman and Virginia Norman, *Creation Today*, season 4, episode 12, "Are Science and Religion Compatible?," https://tinyurl.com/2p9bvry2.

17. Gary Bates, "Is Evolution Really Just about Science?,"

Creation Ministries International, December, 2020, https://tinyurl.com/yc3ew46e.

18. M. Elizabeth Barnes et al., "Accepting Evolution Means You Can't Believe in God: Atheistic Perceptions of Evolution among College Biology Students," *CBE-Life Sciences Education* 19, no. 2 (2020): 1–13.

19. Bret Jaspers, "At CPAC in Dallas, Conservative Elites Urge Activists to Stay Motivated against Democrats," KERA, July 12, 2021, https://tinyurl.com/rjxfp55k.

20. Phyllis Schlafly, "Criticism of Evolution Can't Be Silenced," Eagle Forum, August 16, 2006, https://tinyurl.com/bpeekraz.

21. Edward Humes, *Monkey Girl: Evolution, Education, Religion, and the Battle for America's Soul* (New York: HarperCollins, 2007), 143.

22. Neela Banerjee, "Christian Conservatives Turn to Statehouses," *New York Times*, December 13, 2004, https://tinyurl.com/pft79bjk.

23. Ann Coulter, *Godless: The Church of Liberalism* (New York: Three Rivers, 2001), 199.

Chapter 8

1. Read more about Mary Mallon: Nina Strochlic, "Typhoid Mary's Tragic Tale Exposed the Health Impacts of 'Super-Spreaders,'" *National Geographic*, March 17, 2020, https://tinyurl.com/73kbmy57; and Filio Marineli, "Mary Mallon (1869–1938) and the History of Typhoid Fever," *Annals of Gastroenterology* 26, no. 2 (2013): 132–34, https://tinyurl.com/2vusycam.

2. Read one of Mary's letters: Mary Mallon to George Francis O'Neill, North Brother Island, June 1909, in "In Her Own Words," NOVA, https://tinyurl.com/y6y6dbyz.

3. "Ten Major Court Cases about Evolution and Creation-

ism," National Center for Science Education, June 6, 2016, https://tinyurl.com/2p9dtt44.

4. Bob Smietana, "Religious Liberty Is on the Decline in America," Lifeway, March 30, 2016, https://tinyurl.com/2p95ted8.

5. Aaron Earls, "Evangelicals Are Passionate about Politics, but Mostly Open to Opinions of Others," Lifeway, October 23, 2018, https://tinyurl.com/34wta5e7.

6. John Fritze and David Jackson, "Trump Calls to 'Liberate' States Where Protesters Have Demanded Easing Coronavirus Lockdowns," *USA Today*, April 17, 2020, https://tinyurl.com/2db5eb3u.

7. Seren Morris, "North Carolina Woman Starts 'Burn Your Mask' Challenge to Ignite Freedom," *Newsweek*, June 17, 2020, https://tinyurl.com/y5yhn3fz.

8. Hannah Knowles, "Wearing a Star of David, Another Lawmaker Compares Coronavirus Measures to the Holocaust," *Washington Post*, June 30, 2021, https://tinyurl.com/5n8mu4x9.

9. David French, "How Can We Escape the COVID-19 Vaccine Culture Wars?," *Time*, June 8, 2021, https://tinyurl.com/yc3suena.

10. Benjamin Fearnow, "I'm Listening to God, Not the WHO: Pastor Robert Jeffress Rejects Holiday Restrictions," *Newsweek*, November 22, 2020, https://tinyurl.com/yayujbnm.

11. Laura Ingraham, "Bring It On!," YouTube video interview with John MacArthur, September 15, 2020, https://tinyurl.com/56apxa44.

12. Laura Ingraham, "Bring It On!"

13. John MacArthur (@johnmacarthur), "OPEN YOUR CHURCH. Hebrews 10:25; Matthew 16:18," Twitter, October 7, 2020, https://tinyurl.com/huydsdf4.

14. Owen Strachan, "John MacArthur's Epic Stand," *To*

Reenchant the World (blog), January 12, 2021, https://tinyurl.com/2p88vyew.

15. Daniel Antoine, "The Archaeology of 'Plague,'" *Medical History Suppl.* 27 (2008): 101–14, https://tinyurl.com/5n775apw.

16. Sheera Frenkel, Ben Decker, and Davey Alba, "How the 'Plandemic' Movie and Its Falsehoods Spread Widely Online," *New York Times*, May 21, 2020, https://tinyurl.com/32pm7awv.

17. Jay Bhattacharya, Sunetra Gupta, and Martin Kulldorff, *The Great Barrington Declaration*, October 4, 2020, https://tinyurl.com/yj7vydpj.

18. See Nisreen A. Alwan et al., "Scientific Consensus on the COVID-19 Pandemic: We Need to Act Now," *Lancet* 396, no. 10260 (2020): 71–72, https://tinyurl.com/422ajwrh; "WHO Chief Says Herd Immunity Approach to Pandemic 'Unethical,'" *Guardian*, October 12, 2020, https://tinyurl.com/4ne6vrkm; and Robert Hart, "Fauci Attacks Herd Immunity Declaration Embraced by White House as 'Total Nonsense,'" *Forbes*, October 15, 2020, https://tinyurl.com/kz7nfdvt.

19. James Enstrom, "Ending California's Lockdown on Churches Is Compatible with Science and Good Health," *Daily Signal*, October 16, 2020, https://tinyurl.com/3b9tny5u; and Sean Fine, "Manitoba Judge Denies Constitutional Challenge Ban on In-Person Religious Services over COVID-19," *Globe and Mail*, October 21, 2021, https://tinyurl.com/27yc888n.

20. Brittny Mejia, "When Coronavirus Invaded Their Tiny Apartment, Children Desperately Tried to Protect Dad," *Los Angeles Times*, January 29, 2021, https://tinyurl.com/3m36cky5.

21. Mejia, "When Coronavirus Invaded."

22. "U.S. Poor Died at Much Higher Rate from COVID than Rich, Report Says," Reuters, April 4, 2022, https://tinyurl.com/359525pb.

23. Lyman Stone, "Christianity Has Been Handling Epidemics for 2000 years," *Foreign Policy*, March 13, 2020, https://tinyurl.com/2p9bunkz.

24. Tim Diebel, "Masks? Vaccines? Bible Is Clear on Loving Others," nwestiowa.com, October 13, 2021, https://tinyurl.com/yhhruz7k.

Chapter 9

1. Richard Barnett, "Diabetes," *Lancet* 375, no. 9710 (January 16, 2010), https://tinyurl.com/2cu6hncf.

2. Diana C. Cooper, "Dogs Help Discover Insulin," Famous Dogs in History, November 10, 2016, https://tinyurl.com/2hf8k7tf.

3. Ted Cruz (@tedcruz), "And Fauci's NIH was involved in torturing puppies," Twitter, October 25, 2021. https://tinyurl.com/yn7ed3u3.

4. Ron DeSantis, "NIH not only funded dangerous gain-of-function research in Wuhan, but it also funded cruel experiments on puppies," Facebook, October 25, 2021, https://tinyurl.com/2u5tmnjp.

5. Molly Roberts, "Anthony Fauci Built a Truce. Trump Is Destroying It," *Washington Post*, July 16, 2020, https://tinyurl.com/2p8cnjme.

6. Margaret Sullivan, "Only in Our Anti-Truth Hellscape Could Anthony Fauci Become a Supervillain," *Washington Post*, June 10, 2021, https://tinyurl.com/vmh3svnv.

7. Terrance Egolf and Rachel Santopietro, *Earth Science*, 4th ed. (Greenville: Bob Jones University Press, 2012), ix.

8. Brian Thomas, "Do We Always Believe What Scientists Say?," Institute for Creation Research, July 16, 2014, https://tinyurl.com/z889xaa9.

9. David Roach, "Creationist Critiques Secular Science on Mohler Radio Program," Southern Baptist Theological Seminary, March 2, 2007, https://tinyurl.com/yma5xuat.

10. John Baumgardner, "Exploring the Limitations of the Scientific Method," Institute for Creation Research, March 1, 2008, https://tinyurl.com/bde2yzcp.

11. Brian Thomas, "Cherry-Picked Age for Key Evolutionary Fossil," Institute for Creation Research, June 8, 2017, https://tinyurl.com/bdh8jd93.

12. Jonathan Sarfati, "Archaeoraptor—Phony 'Feathered' Fossil," Creation Ministries International, February 3, 2000, https://tinyurl.com/33fc65ay.

13. Gary Parker, "Vertebrates: Animals with Backbones," Answers in Genesis, March 28, 2016, https://tinyurl.com/54t777b5.

14. J. W. Wartick, "Why I Left the Lutheran Church–Missouri Synod: Points of Fracture (Part 1): Science as a Child," *Reconstructing Faith* (blog), March 21, 2022, https://tinyurl.com/yb69evjm.

15. Gary Parker, "The Fossil Evidence," Answers in Genesis, March 28, 2016, https://tinyurl.com/2p9fw4yt.

16. Thomas, "Cherry-Picked Age."

17. David Menton, "Ape-Woman Statue Misleads Public: Anatomy Professor," *Creation* 19, no. 1 (1996), https://tinyurl.com/3xyj6pcn.

18. "A Science Kit about Science," Parent Company (an online resource for homeschooled children), accessed July 10, 2022, https://tinyurl.com/yudp2wms.

19. John G. West, "The Tragedy of Francis Collins's Model for Science-Faith Integration," Evolution News, October 18, 2021, https://tinyurl.com/abn78822.

20. Dhruv Khullar, "Faith, Science, and Francis Collins," *New Yorker*, April 7, 2022, https://tinyurl.com/2p85tu26.

21. Rand Paul (@RandPaul), Twitter, November 28, 2021, https://tinyurl.com/ykxp3p4f.

22. Rodney Kennedy, "The Same Old Evangelical Enemies," *Word and Way*, January 12, 2022, https://tinyurl.com/5n73up3c.

23. Jason Lemon, "Pastor Says Fauci Should Be Waterboarded until He Admits to Working with China to Create COVID," *Newsweek*, April 12, 2021, https://tinyurl.com/yck2brvz.

24. Joel Webbon, "How Dr. Fauci Successfully Deceived the Whole World," Right Response Ministries, January 19, 2022, https://tinyurl.com/yc6v9sut.

Chapter 10

1. Chris Tomlin, Jesse Reeves, Jonas Myrin, and Matt Redman, "Our God," sung by Chris Tomlin on *And if Our God Is for Us . . .* (Capitol Christian Music Group), 2010, track 1, accessed November 17, 2022, https://tinyurl.com/v9z4efax.

2. "Years of Living Dangerously Premier Full Episode," April 6, 2014, in *The Years Project*, produced by James Cameron, Jerry Weintraub, and Arnold Schwarzenegger, YouTube video, 58:39, https://tinyurl.com/sdsbs4wv.

3. "Years of Living Dangerously."

4. John Cook et al., "Consensus on Consensus: A Synthesis of Consensus Estimates on Human-Caused Global Warming," *Environmental Research Letters* 11, no. 4 (2016), 048002, https://tinyurl.com/4bfk9tkz; Katharine Hayhoe, "Climate in the Time of COVID," BioLogos, August 20, 2020, https://tinyurl.com/mwjxaa55; and US Global Change Research Program, *Fourth National Climate Assessment*, vol. 2, 2018, https://tinyurl.com/au5xy2sk.

5. Katharine Hayhoe, "Climate Science 101," Covering Climate Now, March 28, 2021, https://tinyurl.com/pv6d8rv8.

6. "What Is Climate Change?" United Nations Climate Action, https://tinyurl.com/bdzy36th.

7. "Climate Change Widespread, Rapid, and Intensifying," Intergovernmental Panel on Climate Change, August 9, 2021, https://tinyurl.com/365b3yrr.

8. Read more about the famous "hockey stick chart": Chris Mooney, "The Hockey Stick: The Most Controversial Chart in Science, Explained," *Atlantic*, May 10, 2013, https://tinyurl.com/mrxmu993.

9. Lisa Falkenburg, "Education Board's Ideas on Warming Raise Tempers," *Houston Chronicle*, April 7, 2009, https://tinyurl.com/bddufjks.

10. John James Kirkwood and Kristine Johnson, "Interview with Jessica Moerman: Climate Change and Christianity," *The Sparrow's Call*, YouTube video, 1:26:28, October 24, 2021, https://tinyurl.com/y4uvtafm.

11. Bob Allen, "Pro-Trump Pastor Robert Jeffress Uses Bible to Debunk Science of Climate Change," Baptist News Global, September 27, 2019, https://tinyurl.com/yc8x95te.

12. Katharine Hayhoe, "Global Weirding with Katharine Hayhoe," Facebook, April 4, 2021. https://tinyurl.com/5a2s3ffa.

13. See Jake Hebert, "Climate Alarmism and the Age of the Earth," Institute for Creation Research, March 29, 2019, https://tinyurl.com/mmbbfyen; and "Global Warming: Fact or Fiction?," Answers in Genesis, https://tinyurl.com/24sw4y9w.

14. Ken Ham, "What or Who Is Causing Climate Change? That Depends on Your Starting Point," Answers in Genesis, July 30, 2020, https://tinyurl.com/4rpwk5cz.

15. Jake Hebert, "The Bitter Harvest of Evolutionary Thinking," Institute for Creation Research, March 31, 2014, https://tinyurl.com/yeyk78k2.

16. "Years of Living Dangerously Premier Full Episode," April 6, 2014, in *The Years Project*, produced by James Cam-

eron, Jerry Weintraub, and Arnold Schwarzenegger, YouTube video, 58:39, https://tinyurl.com/sdsbs4wv.

17. Akos Balogh, "Why Christians Should Avoid Alarmism (and What to Do Instead)," Gospel Coalition, October 25, 2021, https://tinyurl.com/3efsafbt.

18. A. P. Carter, "I Can't Feel At Home Any More," 1931, https://tinyurl.com/yvh95m7c.

19. John MacArthur, "The End of the Universe, Part 2," Grace to You, September 28, 2008, https://tinyurl.com/rvp4zmck.

20. K. L. Marshall, "Revisiting the Scopes Trial: Young-Earth Creationism, Creation Science, and the Evangelical Denial of Climate Change," *Religions* 12, no. 2 (2021): 133, https://tinyurl.com/nuhztu8e.

21. Marshall, "Revisiting the Scopes Trial," 133.

22. Jason M. Breslow, "Bob Inglis: Climate Change and the Republican Party," *Frontline*, October 23, 2012, https://tinyurl.com/4kkdvht9.

23. Bill Moyers, interview with Katharine Hayhoe, "Climate Change—Faith and Fact," *Moyers and Company* blog), 25:16, September 12, 2014, https://tinyurl.com/yj3d6juf; and Katharine Hayhoe, interviewed by Jim Stump, "Climate in the Time of COVID," March 12, 2020, in *Language of God*, podcast, https://tinyurl.com/mwjxaa55.

24. Evan Lehmann, "Conservatives Lose Faith in Science over Last 40 Years," *Scientific American*, March 30, 2012, https://tinyurl.com/mvrpa7zv.

25. Paul Douglas (keynote presentation, Jesus Rode a Dinosaur: Talking to Youth about Faith in a Scientific Age, Science and Youth Ministry Capstone Event, Edina, MN, May 7–8, 2018).

26. "Road to the White House," C-Span, video, 43:11, August 17, 2011, https://tinyurl.com/wa7sz345.

27. Sarah Posner, "The Real Reasons Why Evangelical

Embrace of Environmentalism Lags," Religion Dispatches, June 24, 2011, https://tinyurl.com/68kebjuh.

28. Marshall, "Revisiting the Scopes Trial," 133.

29. John MacArthur, "End of the Universe."

30. "Force of Nature," *Throughline*, NPR, April 22, 2021, radio broadcast, 51:44, https://tinyurl.com/bdfmvfk7.

31. Christiane Amanpour, interview with Katharine Hayhoe, "We Must Talk Solutions to Climate Change," *Amanpour and Co.*, January 8, 2019, https://tinyurl.com/n9u7xp58.

32. Lauri Lebo, "Creationism and Global Warming Denial: Anti-Science's Kissing Cousins?," Religion Dispatches, March 18, 2010, https://tinyurl.com/2v2jbjn9.

33. Michaela Melia, "School Lessons Targeted by Climate Change Doubters," AP News, March 6, 2019, https://tinyurl.com/m3e4my75.

34. Elizabeth Harball, "Teach the Controversy Comes to Climate Science," *Scientific American*, March 6, 2013, https://tinyurl.com/2wcy6rd4; and Leslie Kaufman, "Darwin Foes Add Warming to Targets," *New York Times*, March 3, 2010, https://tinyurl.com/3b8z73ap.

35. Harball, "Teach the Controversy."

36. Harball, "Teach the Controversy."

37. Jake Hebert, "Genesis and Climate Change," Institute for Creation Research, August 11, 2016, https://tinyurl.com/y8y28exh.

Chapter 11

1. Ellen Callaway, "Dolly at 20: The Inside Story on the World's Most Famous Sheep," *Nature*, June 29, 2016, https://tinyurl.com/yzfdtupd.

2. Karen Weintraub, "20 Years after Dolly the Sheep Led the Way—Where Is Cloning Now?," *Scientific American*, July 5, 2016, https://tinyurl.com/yxhx4stz.

3. Monya Baker, “Stem Cells Made from Cloned Human Embryos,” *Nature*, April 28, 2014, https://tinyurl.com/mupj26yw.

4. Snigdha Prakash and Vikki Valentine, “Timeline: The Rise and Fall of Vioxx,” NPR, November 10, 2007, https://tinyurl.com/mpduj4dz.

5. Anna MacDonald, “Stem Cells in Drug Discovery,” Cell Science, April 4, 2017, https://tinyurl.com/4k8e9xwx.

6. “Why We Need Stem Cell Avatars: Susan Solomon Speaks at IGNITION Conference,” New York Stem Cell Foundation, December 11, 2018, https://tinyurl.com/39562c7; and TED, “Susan Solomon: The Promise of Research with Stem Cells,” YouTube video, 14:58, September 13, 2012, https://tinyurl.com/yssz6bw6.

7. Baker, “Stem Cells Made”, and Geeta Shroff, “Human Embryonic Cell Therapy in Chronic Spinal Cord Injury: A Retrospective Study,” *Clinical and Translational Science* 9, no. 3 (2016): 168–75, https://tinyurl.com/mr4cy975.

8. Priyanka Runwal, “Here Are the Facts about Fetal Cell Lines and COVID-19 Vaccines,” *National Geographic*, November 19, 2021, https://tinyurl.com/bddnh983.

9. “Johnson & Johnson’s Janssen COVID-19 Vaccine,” CDC, August 3, 2022, https://tinyurl.com/mrx4n6jk.

10. “COVID-19 Vaccines and Fetal Cell Lines,” North Dakota Health, December 1, 2021, https://tinyurl.com/mw5m6nsf.

11. “COVID-19 Vaccines and Fetal Cell Lines.”

12. Meredith Wadman, “Abortion Opponents Protest COVID-19 Vaccines’ Use of Fetal Cells,” *Science*, June 5, 2020, https://tinyurl.com/7nk2ttrx.

13. Runwal, “Facts about Fetal Cell Lines.”

14. “Human Fetal Tissue: A Critical Resource for Biomedical Research,” International Society for Stem Cell Research, https://tinyurl.com/mr2xu4p3.

15. Varnee Murugan, "Embryonic Stem Cell Research: A Decade of Debate from Bush to Obama," *Yale Journal of Biology and Medicine* 82, no. 3 (2009): 101–3, https://tinyurl.com/2p865rde.

16. Jocelyn Kaiser and Meredith Wadman, "Trump Administration Releases Details on Fetal Tissue Restrictions," *Science*, July 26, 2019, https://tinyurl.com/2p8mjnw8.

17. Weintraub, "Where Is Cloning Now?"

18. Sasha Pezenik and Anne Flaherty, "Conservatives Confront Moral Dilemma of Vaccines and Treatments Derived from Fetal Tissue Cells," ABC News, October 18, 2020, https://tinyurl.com/2xars4d7.

19. Meredith Wadman and Jocelyn Kaiser, "NIH Chief Defends Use of Human Fetal Tissue as Opponents Decry It before Congress," *Science*, December 13, 2018, https://tinyurl.com/2p9c5a9m.

20. "On COVID Vaccinations, Pope Says Health Care Is a 'Moral Obligation,'" NPR, January 10, 2022, https://tinyurl.com/edtmp6n2.

21. Curtis Chang and David French, interview by Jim Stump, "Christians and the Vaccine," BioLogos, podcast, 1:02:48, March 18, 2021, https://tinyurl.com/28s88t3n.

22. Chang and French, "Christians and the Vaccine."

23. Wadman and Kaiser, "NIH Chief Defends Use."

24. Katie Yoder, "Report: Top Biden Science Advisor Defended Using Aborted Baby Parts in Research while NIH Director," *National Catholic Register*, March 11, 2022, https://tinyurl.com/4j78257f.

25. Nidhi Subbaraman, "Science Misinformation Alarms Francis Collins as He Leaves Top NIH Job," *Nature*, December 3, 2021, https://tinyurl.com/ykabudkr.

26. Megan Basham, "How the Federal Government Used Evangelical Leaders to Spread Covid Propaganda to Churches," *Daily Wire*, February 1, 2022, https://tinyurl.

com/36vbhv5m; and Alice Miranda Ollstein, "Anti-Abortion Groups Demand Ouster of NIH Chief over Fetal Tissue," *Politico*, December 18, 2018, https://tinyurl.com/4ezeyj69.

27. See, for example, the interview with Megan Basham, "Christian Leaders Complicit in Covid Propaganda Machine—Part 2," Christian Worldview, February 26, 2022, https://tinyurl.com/mry2kh9k; and John G. West, "The Tragedy of Francis Collins's Model for Science-Faith Integration," Evolution News, October 18, 2021, https://tinyurl.com/abn78822; and John Zmirak, "Francis Collins Shows Us: Theistic Evolution Rots the Brain," Stream, October 11, 2021, https://tinyurl.com/ymcxx45c.

28. Heidi Ledford and Ewen Callaway, "Pioneers of Revolutionary CRISPR Gene Editing Win Chemistry Nobel," *Nature*, October 7, 2020, https://tinyurl.com/bdr6nzj5.

29. For more about Victoria's game-changing treatment, see Rob Stein, "First Sickle Cell Patient Treated with CRISPR Gene-Editing Still Thriving," NPR, December 31, 2021, https://tinyurl.com/3rts362m; and Stein, "A Year in, 1st Patient to Get Gene Editing for Sickle Cell Disease Is Thriving," NPR, June 23, 2020, https://tinyurl.com/5f3w2xyp.

30. Stein, "Gene Editing for Sickle Cell Disease."

31. Jessica Hamzelou, "Exclusive: World's First Baby Born with New '3 Parent' Technique," New Scientist, September 27, 2016, https://tinyurl.com/2p8u2556.

32. "Philippines Approves Commercial Use of Genetically Engineered Rice," Reuters, August 25, 2021, https://tinyurl.com/nb4n3vhf.

Chapter 12

1. David Cyranoski, "What CRISPR-Baby Prison Sentences Mean for Research," *Nature*, January 3, 2020, https://tinyurl.com/32nmkzby. Following his release from prison,

Dr. He admitted he "acted too quickly" but has yet to express regret. Hannah Devlin, "Scientist Who Edited Babies' Genes Says He Acted Too Quickly," *Guardian*, February 4, 2023, https://tinyurl.com/2p9yyc57.

2. Peter Wehner, "The Moral Universe of Timothy Keller," *Trinity Forum*, December 5, 2019, https://tinyurl.com/bdceh68b.

3. NRA (@NRA), "Someone should tell self-important anti-gun doctors to stay in their lane," Twitter, November 7, 2018, https://tinyurl.com/yc5rekr3.

4. Matthew Haag, "Doctors Revolt after N. R. A. Tells Them to 'Stay in Their Lane' on Gun Policy," *New York Times*, November 13, 2018, https://tinyurl.com/34b6h4a8.

5. Sean B. Carroll, "The Denialist Playbook," *Scientific American*, November 8, 2020, https://tinyurl.com/4cbk92mk.

6. Mark A. Noll, *The Scandal of the Evangelical Mind* (Grand Rapids: Eerdmans, 1994), 196.

7. John James Kirkwood and Kristine Johnson, "Interview with Jessica Moerman: Climate Change and Christianity," *The Sparrow's Call*, YouTube video, 1:26:28, October 24, 2021, https://tinyurl.com/y4uvtafm.

8. John Rennie and Steve Mirsky, "Six Things in *Expelled* That Ben Stein Doesn't Want You to Know," *Scientific American*, April 16, 2008, https://tinyurl.com/352m4k2a.

9. Finn R. Pond and Jean L. Pond, "Scientific Authority in the Creation-Evolution Debates," *Evolution: Education and Outreach* 3 (2010): 641–60, https://tinyurl.com/4u6avahk.

10. Kirkwood and Johnson, "Interview with Jessica Moerman."

11. "Six Reasons Young Christians Leave Church," Barna, September 27, 2011, https://tinyurl.com/zbxeb8pd.

12. Yonat Shimron, "Black Protestants Aren't Least Likely

to Get a Vaccine; White Evangelicals Are," Religion News Service, March 5, 2021, https://tinyurl.com/yckru4m9.

13. Mike Cummings, "Unvaccinated White Evangelicals Appear Immune to Pro-Vaccine Messaging," *Yale News*, November 29, 2021, https://tinyurl.com/5n8w5cvu.

14. "U.S. Poor Died at Much Higher Rate from COVID Than Rich, Report Says," Reuters, April 4, 2022, https://tinyurl.com/359525pb.

15. Jim Stump, interview with Sandra Richter, "We Are Only Renters Here," BioLogos, April 22, 2021, https://tinyurl.com/246ammdm.

16. Beth Moore (@BethMooreLPM), "Stare in the face of what some of you are saying: MY RIGHTS ARE MORE IMPORTANT THAN YOUR LIFE. SORRY, NOT SORRY. If you're not going to get vaccinated, for the love of God, PUT ON A MASK IN PUBLIC PLACES WITH VULNERABLE PEOPLE. Go ahead & unfollow me. I don't care. FOLLOW JESUS," Twitter, August 10, 2021, https://tinyurl.com/2mnanfweu; Greg Locke (@ pastorlocke), "You are pathetic. A mouth piece for virtue signaling to an ungodly culture is what you've become. You lecture Believers and stand with Communists. Sit down!!!" Twitter, August 11, 2021, https://tinyurl.com/4r4k4x3k.

17. Jonathan Rauch, *The Constitution of Knowledge: A Defense of Truth* (Washington, DC: Brookings Institution Press, 2021), 44.

18. Charles Taylor, *A Secular Age* (Cambridge, MA: Harvard University Press, 2007), 25.

19. Noll, *Scandal of the Evangelical Mind*, 7.

Index